Modern Design in Plastics

MODERN DESIGN IN PLASTICS

D.P. Greenwood

John Murray

First published 1983
by John Murray (Publishers) Ltd
50 Albemarle Street, London W1X 4BD

Typeset in Great Britain by
Servis Filmsetting Ltd, Manchester.
Printed by Camelot.

British Library Cataloguing in Publication Data

Greenwood, D.P.
Modern Design in Plastics.
1. Plastics—Moulding
I. Title
668.4'12 TP1150

ISBN 0-7195-3966-8

CONTENTS

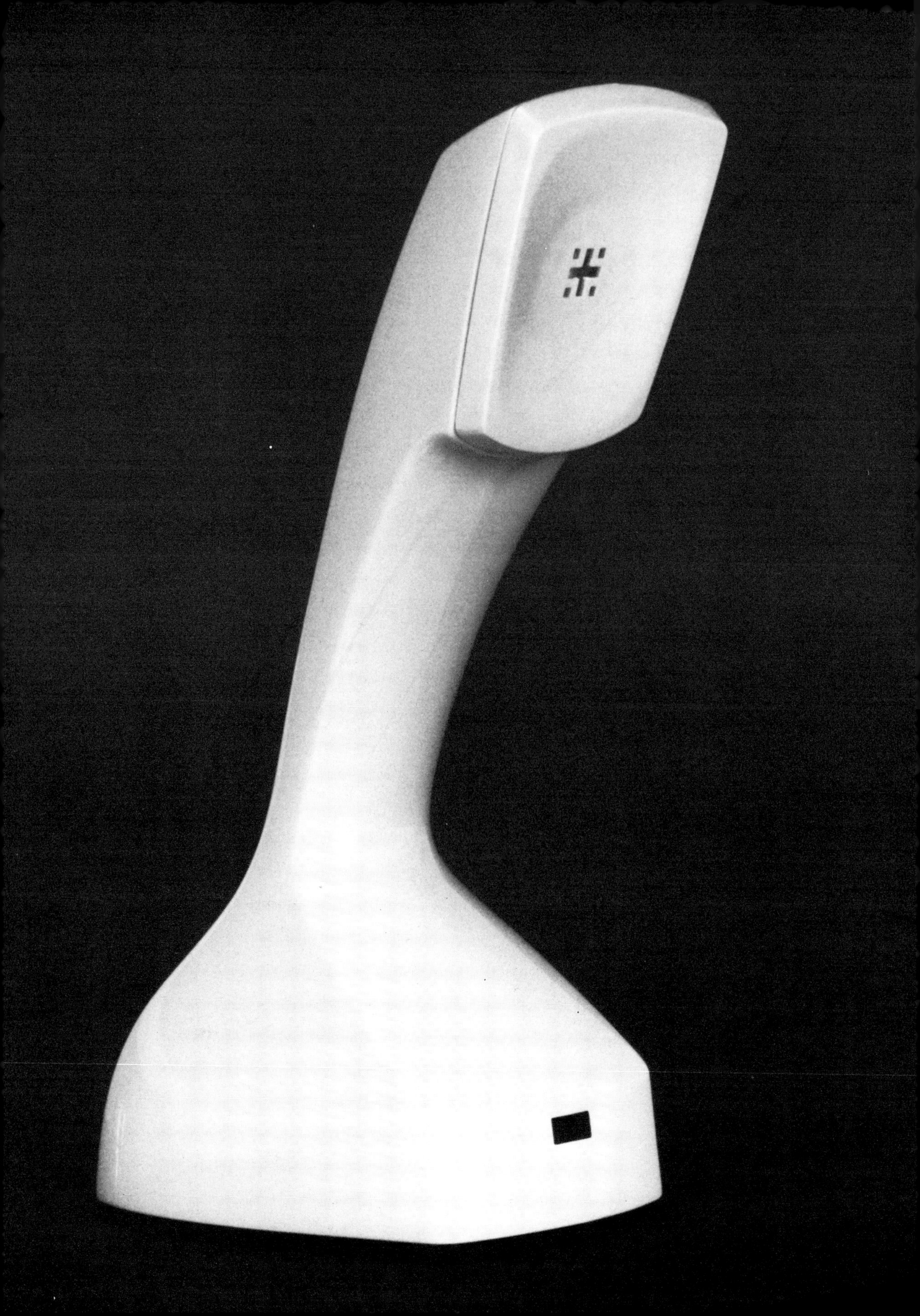

INTRODUCTION

Modern plastics are almost entirely a product of human ingenuity and technical skill. In terms of function, they can often do jobs that are simply beyond the capabilities of other materials: they can be made into pipes that don't burst, tanks that don't corrode, boats that fold up but are almost impossible to sink, chairs that give but never break, and shoe-soles that last practically for ever. Plastics are also remarkably versatile from the manufacturing point of view. A handle that fits the grip perfectly can be moulded in seconds, whereas a piece of wood would have to be cut, shaped, sanded and polished. Yet at the same time, plastics can be beautiful. Soft or hard, flexible or rigid, smooth or textured, transparent or any colour of the rainbow, plastics are perhaps more than any other material exactly what the designer makes them.

Good design has been defined as the marriage of function and form, and the examples chosen for this book are arranged to show how these two elements combine. In the earlier sections the emphasis is on function, with examples from the fields of industry, engineering and construction, showing that in the real world good design must often be severely practical. In addition to the assembled items, photographs of the component parts are often included, to show how the product works and some of the problems involved in designing the whole piece. As the book develops, more purely aesthetic considerations come into play. The wide selection of articles for leisure, sport and games shows how things designed to serve specific purposes can also look exciting, and visual appeal plays an even more important part in the furnishings and household items that follow. The book ends with a section on jewellery and sculpture, where aesthetic considerations are all-important. All the photographs have been carefully chosen to illustrate a wide range of modern plastics from the mass-production of everyday items to the crafting of unique pieces. Each photograph is fully captioned, with the materials specified according to their basic types (sometimes by trade names). The collection as a whole shows a very wide variety of products, many of which have already been accepted as works of excellence as regards both utility and appearance.

ACKNOWLEDGMENTS

More than sixty companies are represented in this book, many of them well-known internationals, others small but equally important in their own specialised fields. The author acknowledges the generous help and interest of these organisations; without their photographs and technical information this book would not have been possible.

With the growing emphasis on craft, design and technology in education, the book offers both teachers and students a source of reference and a wide range of ideas. The hand-crafted pieces have an obvious relevance to workshop activities in schools and colleges, but the mass-produced items can also be used as a source of inspiration for student projects. These items can often be produced on a one-off basis, using a variety of materials and a range of hand or machine skills. Some such items will require the use of moulds or jigs to achieve a satisfactory result, and this in turn introduces the skills needed to obtain the appropriate finish. But it is left very much to the reader to extract and develop these ideas.

This book takes its place alongside Richard Stewart's *Modern Design in Wood* and *Modern Design in Metal*, which have become standard reference books in schools and colleges. It is the first such book to deal specifically with the wide range of modern plastics.

INDUSTRY & ENGINEERING

ADVANCED PASSENGER TRAIN

The cab shell is part GRP and part aluminium. Made by British Rail Engineering Ltd.

a **GEAR WHEELS**

Various wheels for the automotive industry made from nylon. Courtesy Akzo Plastics.

b **RADIATOR CAPS**

Moulded in 'Akulon' Courtesy Akzo Plastics.

c **WORM-WHEELS**

Used in speedometers. Also in 'Akulon'. Courtesy Akzo Plastics.

d **FILTER**

Hydraulic circuit filter for brake and clutch systems, made in nylon. Courtesy Akzo Plastics.

e **DOMESTIC WATER FILTER**

The body and nearly all components are moulded in acetal copolymer. The outlet arm is chrome-plated ABS. Made by Doulton Industrial Products, Courtesy Amcel Ltd.

a

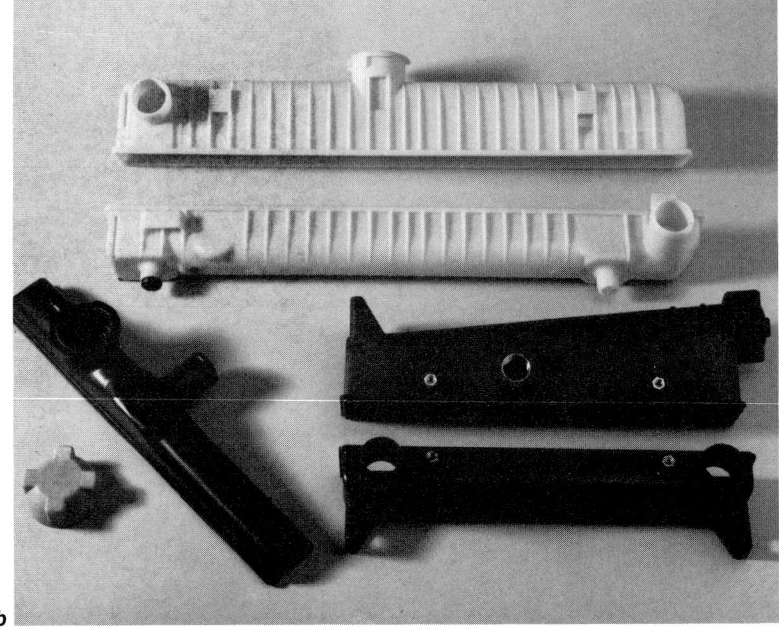

b

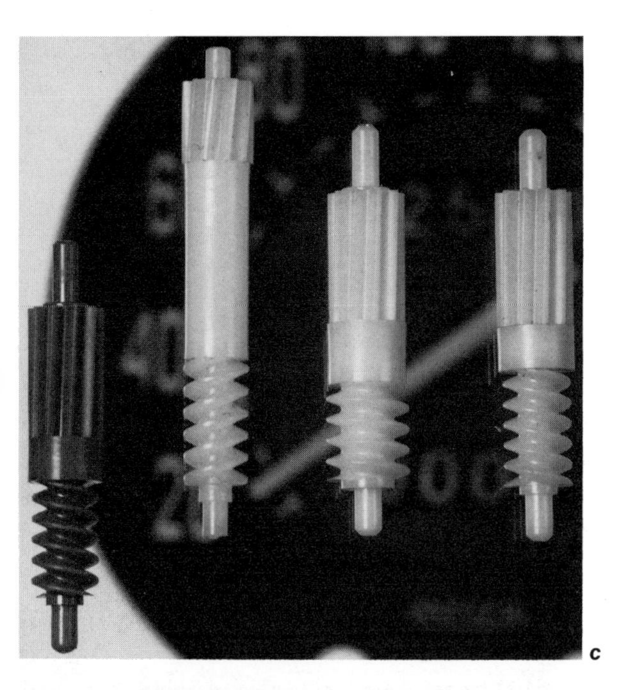

c

d

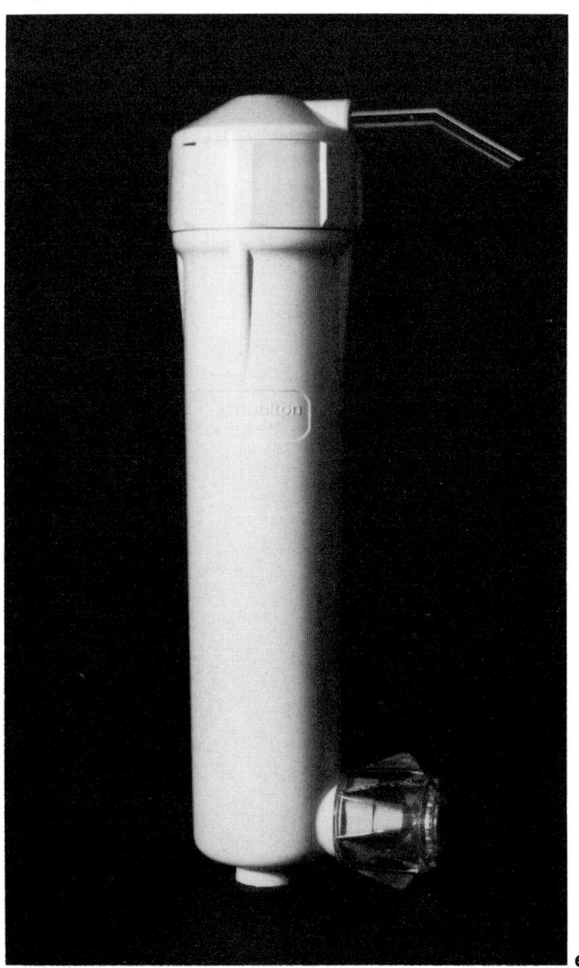

e

a **CAR GRILLE**

Moulded in a special grade of Monsanto's ABS to give rigidity, toughness and heat resistance. Made by Pressed Steel Fisher.

b **TRAIN**

Outer shells for the high-speed train made in GRP. Courtesy British Rail.

c **AIRSHIP**

All the windows are shaped from 'Perspex', ICI's acrylic sheet. Made by Suntex Ltd.

a

b

c

d

e

d **TORNADO AIRCRAFT FACSIMILE**

A full-size replica complete in every detail, produced for the RAF recruiting organisation. Made by Specialised Mouldings Ltd.

e **COMMERCIAL VEHICLE CAB**

The cab is built from 23 individual press-moulded GRP mouldings. The air deflector on the roof is also press-moulded. Courtesy Fibreglass Ltd.

a **CHEMICAL SILOS**

The two GRP hoppers are designed to hold chemicals. They are being located on a platform enabling tankers to run beneath for loading purposes. Courtesy Fibreglass Ltd.

b **CROP SPRAYER**

The tank is rotationally moulded in polyethylene by Argoe Plastics Ltd.

a

b

c **BLIND MECHANISM**

Winding system made of nylon for venetian blinds. Courtesy Akzo Plastics.

d **CONVEYOR CHAINS**

Designed for materials handling and widely used in the food and building brick industries. The chains are made by Ling Systems Ltd in acetal copolymer. Courtesy Amcel Ltd.

e **MINI-BANDSAW**

The top and bottom housings, their doors and the motor cowl are moulded in nylon reinforced with glass bead to minimise warpage. The switch plates are reinforced with glass fibre, which has good mechanical and thermal properties. All components moulded by Britton Plastics for Burgess Power Tools. Courtesy ICI Plastics Division.

c

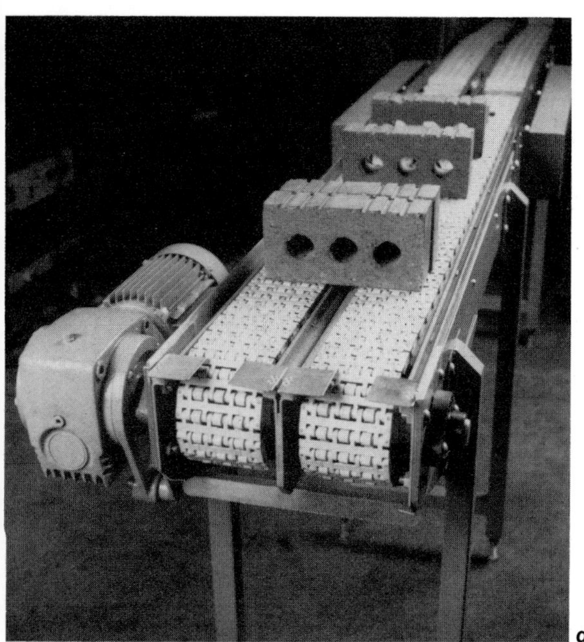

d

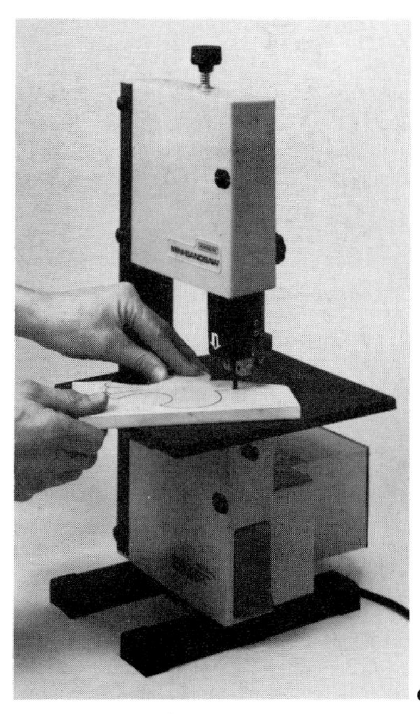

e

a **SAFETY GOGGLES**

Moulded in 'Welvic', ICI's PVC. Made by Superguard Ltd.

b **EYE GUARDS**

The lenses of these safety spectacles are made from 'Rocel' polished acetate sheet. Courtesy Courtaulds Acetate Ltd.

c **VISOR**

Polished cellulose acetate sheet is used for the visor. Courtesy Courtaulds Acetate Ltd.

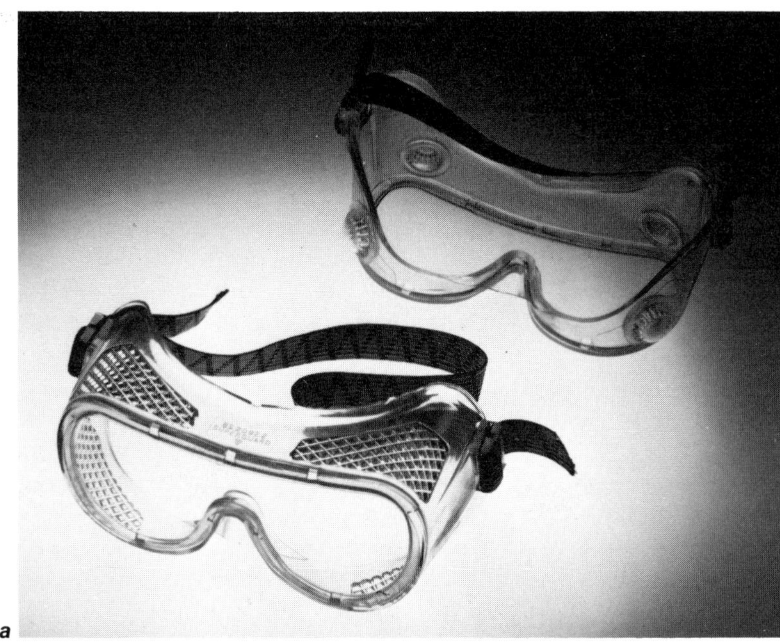

a

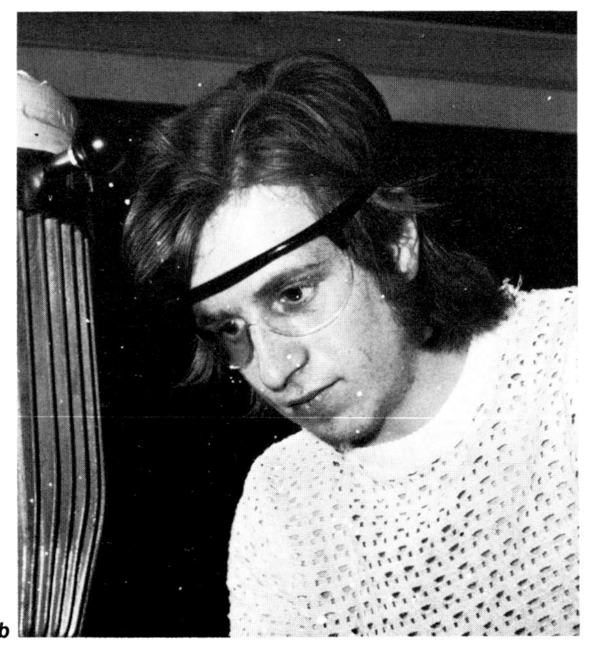

b

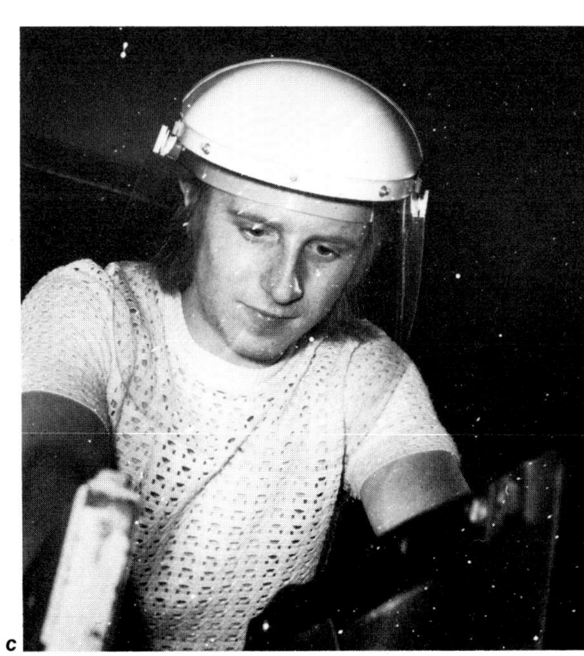

c

d

_d_ ELECTROPLATING BARRELS

Fabricated in polypropylene by Varicol Ltd. Courtesy Courtaulds Acetate Ltd.

_e_ ELECTROPLATING JIG

Fabricated in 'Polyplex', Courtaulds Acetate's extruded polypropylene sheet. Made by Varicol Ltd.

_f_ ELECTROPLATING TANKS

The two outer tanks contain water washes and the inner tank zinc coating solution. Made of polypropylene by Varicol Ltd. Courtesy Courtaulds Acetate Ltd.

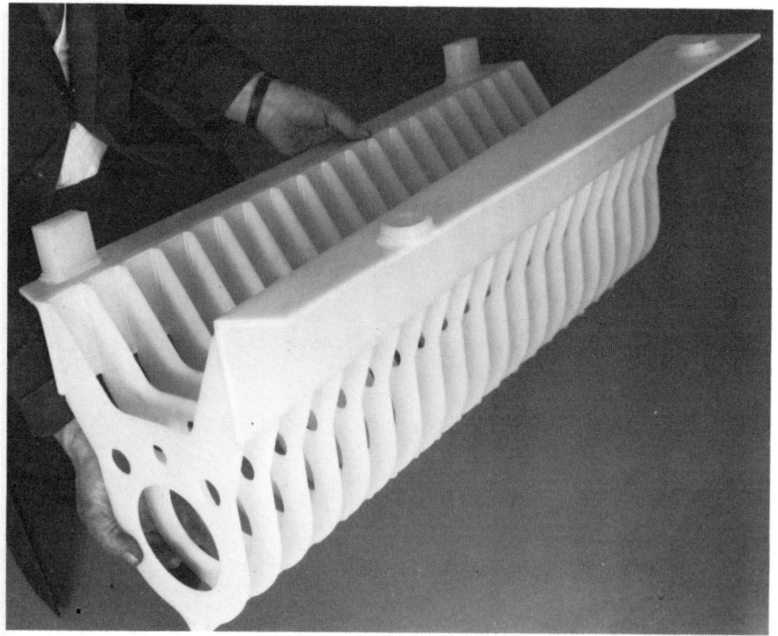

e

f

a **LAMP HOUSINGS**

Used in microfilm reading. They are moulded in polyethersulphone, which withstands high operating temperatures and can be metallised. Made in Switzerland by WEZ. Courtesy ICI Plastics Division.

b **LENS FILTER HOLDER**

A hinge integral with the lens holder is forced flat against the base and then ultrasonically welded. Two pairs of springs are moulded into each of the three filter slides, providing a strong grip to retain the lenses securely. Made in acetal copolymer by Downey Plastics Ltd. Courtesy Amcel Ltd.

c **MICROFILM PROJECTOR**

The grille/light baffle is made from 'Victrex', ICI's polyethersulphone. Made by Microfilm Communications.

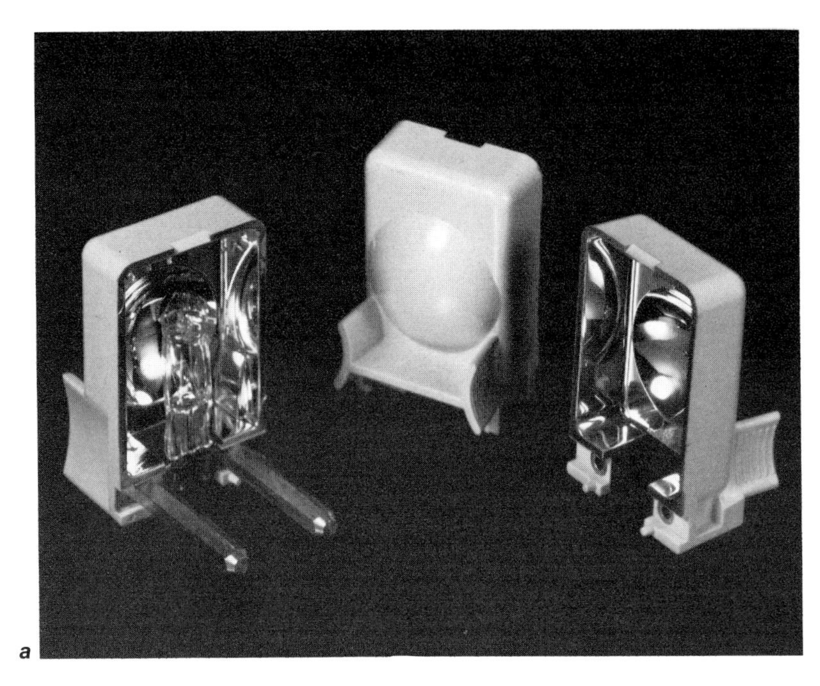

a

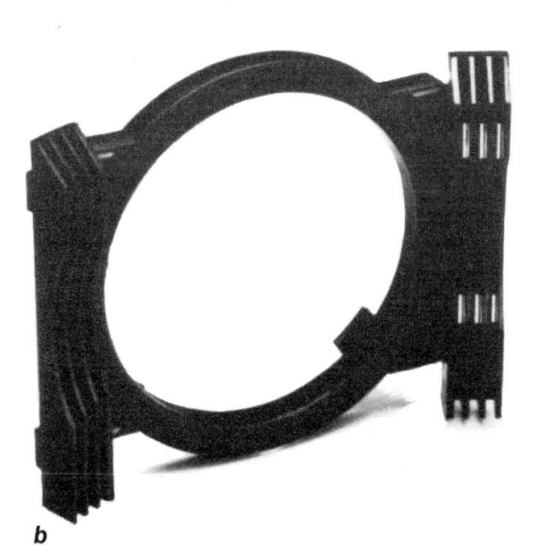

b

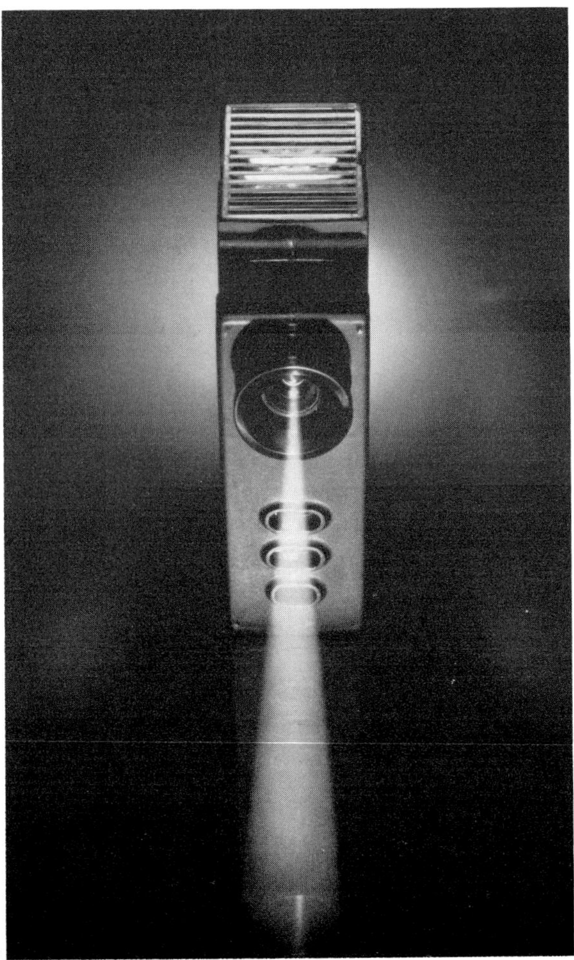

c

BUILDING ACCESSORIES

ROAD SURFACE BOX

Made from a tough outdoor grade of propathene, ICI's polypropylene. The boxes are used by the gas or water authorities. Made by GC Plastics Ltd.

a **DRAINAGE GULLY**

The grating is made from 'Propathene', ICI's polypropylene. The other components are made from 'Welvic', ICI's vinyl compound. Made by G. Molyneux (Products) Ltd.

b **INSPECTION CHAMBERS**

A range of underground drainage chambers injection-moulded in uPVC. The pipe is extruded. Made by Marley Extrusions Ltd.

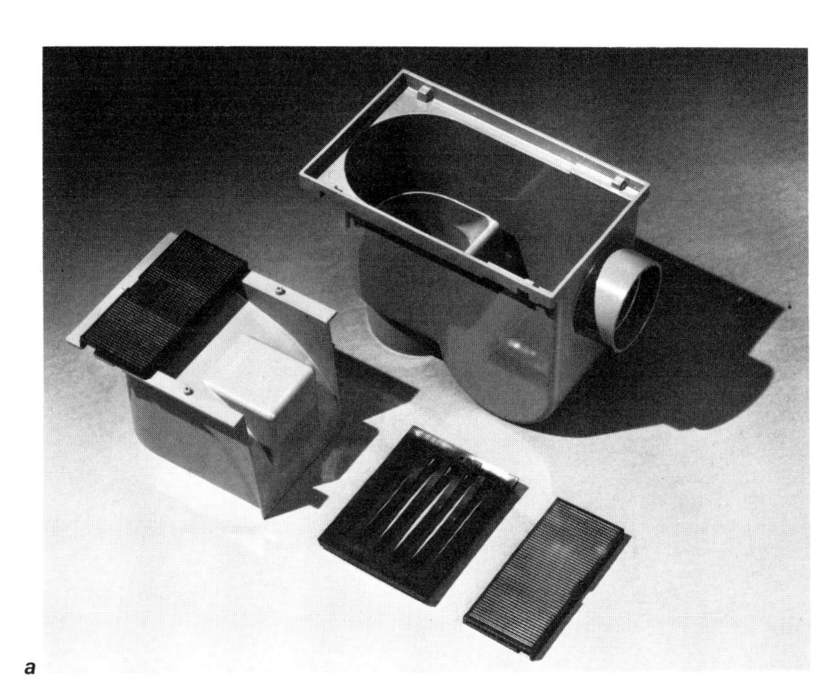

a

b

c

c MIXER TAP COMPONENTS

The components are moulded in 'Kematal' Amcel's acetal copolymer. The end-nut is moulded in a 25 per cent glass-coupled grade of resin. Made by Damixa of Denmark.

d GULLY HOPPER

The main body is made from uPVC. The grating and cut-out plate is of polypropylene. Made by Marley Extrusions Ltd.

e PLUMBING FITTING

One of the Acorn range of fittings made by Bartol for hot and cold water systems. The body is made in polybutylene and incorporates a stainless steel grab-ring. The screw-on end caps are in acetal copolymer. Photo by Amcel Ltd.

d

e

a **SANITARYWARE**

The bath is made from ICI's perspex acrylic sheet by vacuum-forming. Accessory fittings are injection-moulded in ABS. Made by Marley Extrusions.

b **WASHBASIN**

Lightweight acetal copolymer corner basin. Ideal for boats, caravans, mobile homes and the replacement market for domestic homes. Made by Armitage Monobond.

c **VENTILATOR CONTROLLER**

Flush fitting into a wall, three-speed, two-direction fan control. The fascia is moulded in melamine, the body casing in phenolic resin. Used in conjunction with the Universal range of ventilators. Courtesy Vent-Axia Ltd.

a

b

c

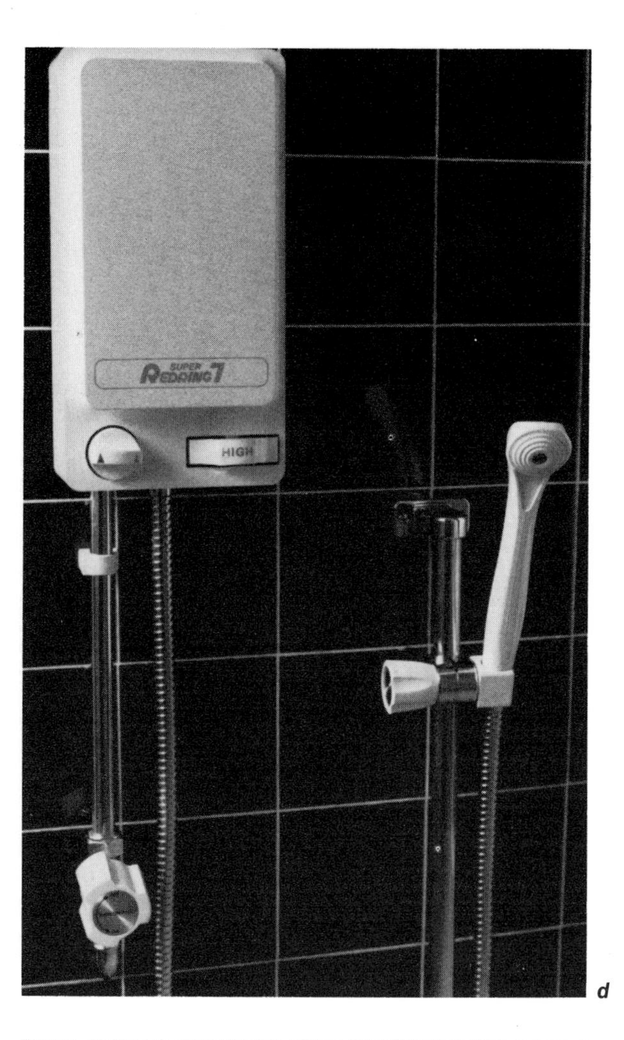

d

d ELECTRIC WATER HEATER

The casing unit and shower head is moulded in styrene terpolymer. Made by Redring Electric.

e UNIVERSAL VENTILATORS

The models are designed to be fitted into wall, roof and window situations. The body housings are of glass-filled nylon, and a large part of the motor is also in glass-filled nylon. Courtesy Vent-Axia Ltd.

f DOMESTIC VENTILATOR

Designed for use in the kitchen, bathroom and utility room, being mounted through a window pane. Its white body, grey grille and black satin finish surround are moulded in ABS. The front and rear mouldings are identical, the grille, being a separate unit enables the angle of the vanes to be reversed for internal and external use. Courtesy Vent-Axia Ltd.

e

f

***a* SECTION OF A TILT-AND-TURN WINDOW**

***b* WINDOW FRAME**

Extruded in PVC by Range Valley Engineering Ltd. Photo by ICI Ltd.

***c* TILT-AND-TURN WINDOW**

Shown in the tilt position, which allows draught-free ventilation.

a

b

c

d

d **SHUTTABLINDS**
The louvres are made from ICI's 'Welvic' PVC, by Telcon Plastics for Guardia Shutters Ltd.

e **SLIDING PATIO DOOR**

f **RESIDENTIAL DOOR**

a, ***c***, ***e***, ***f*** All of these units are made from uPVC, heat-sealed joints being used for their construction. Made by Hepworth Industrial Plastics Ltd.

e

f

a **INDUSTRIAL WAREHOUSE COMPLEX**

The model is on a plywood base covered in polystyrene sheet. The warehouses and boundary wall are also styrene sheet. Roof lights are of cellulose sheet and the trucks are of balsa wood. Designed and the models made by Fletcher, Ross & Hickling.

b **WAREHOUSES**

The model has a perspex base with a cardboard back drop. Buildings and the truck are of polystyrene sheet. Car and fork-lift of perspex. The windows are cellulose sheet. Made by Fletcher, Ross & Hickling.

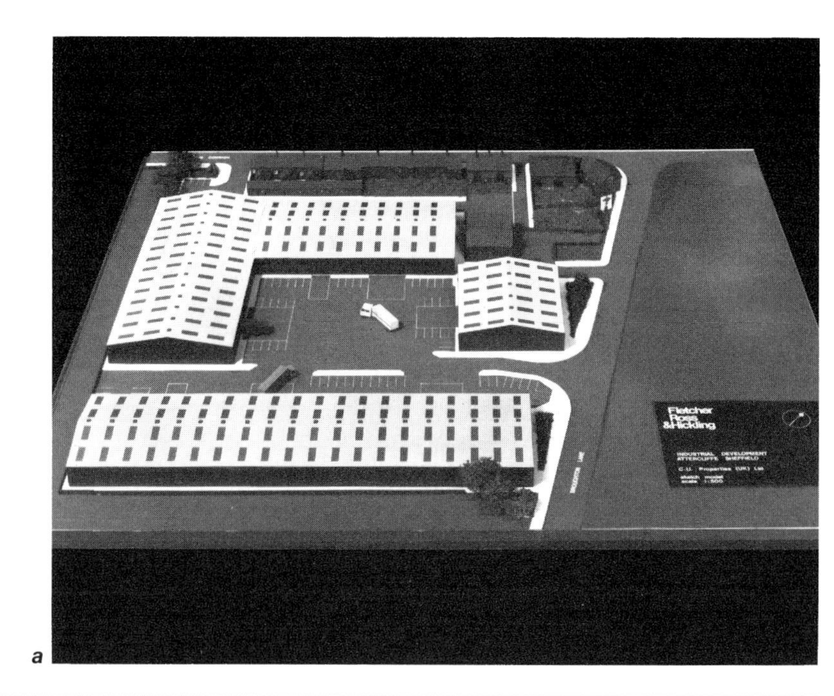

a

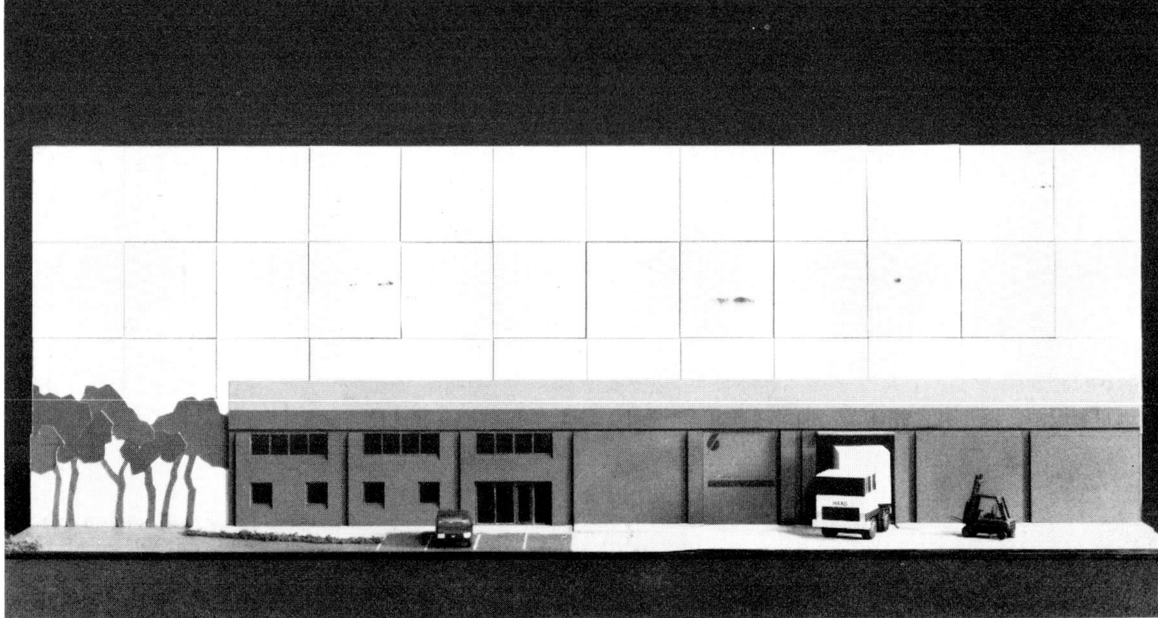

b

c

c, *d* FACTORY SITE

Building structure model is of cardboard covered in cellulose sheet. Roof lights, window and security fence are of acetate sheet. Signs on the trucks and building are Letraset. Street lighting posts, rear storage racking, perimeter fence, main gate barrier and the steps handrail are all of polystyrene rod. The landscape is of expanded styrene covered with plaster. The factory sign is perspex. Designed and produced by Fletcher, Ross & Hickling.

d

a **GARAGE DOOR**

Door-frame of galvanised steel faced with vacuum-formed high-impact ABS. Made by Bluehart Ltd.

b **HOUSING**

The wall cladding system is in extruded rigid PVC. Produced by Telcon Plastics Ltd.

c **PLASTICS DEVELOPMENT HOUSE**

PVC weather boarding is used for the cladding of the house. 'Corvic', ICI's vinyl polymer, is used for the combined fascia/soffit and rainwater gutter. All were made and fitted by Marley Plumbing Co. Ltd. Courtesy ICI Plastics Division.

a

b

c

GARDENING

PATIO TUBS

Moulded in polypropylene by Stewart Plastics.

a **CLOCHES**

Moulded in natural polypropylene, the end-pieces are from clear polystyrene. Made by Stewart Plastics.

b, ***e*** **INTERIOR PLANT CONTAINERS**

Designed and moulded in GRP for Plantdecor Ltd.

a

b

c

d

c PROPAGATOR

The base is vacuum-formed ABS. The frame is aluminium and glazed with acrylic. Manufactured by Jemp Engineering Ltd.

d PROPAGATORS

Polystyrene is used for both tray and cover. Made by Stewart Plastics.

e See previous page.

f SEED TROUGH

The seed troughs and frame are produced in high-impact polystyrene. Manufactured by Skarsten Ltd.

e

f

a **WATER POOL PUMP**

The casing is polypropylene, injection-moulded. Made by Hozelock Ltd.

b **GARDEN LIGHTS**

Injection-moulded using ABS. Available with coloured lenses. Made by Hozelock Ltd.

c **SPRAYER**

Made in polythene, it connects to a water hosepipe. Made by Hozelock Ltd.

d **SEED/FERTILIZER DISTRIBUTOR**

Injection-moulded hopper. Produced by Wolf Tools Ltd.

a

b

c

d

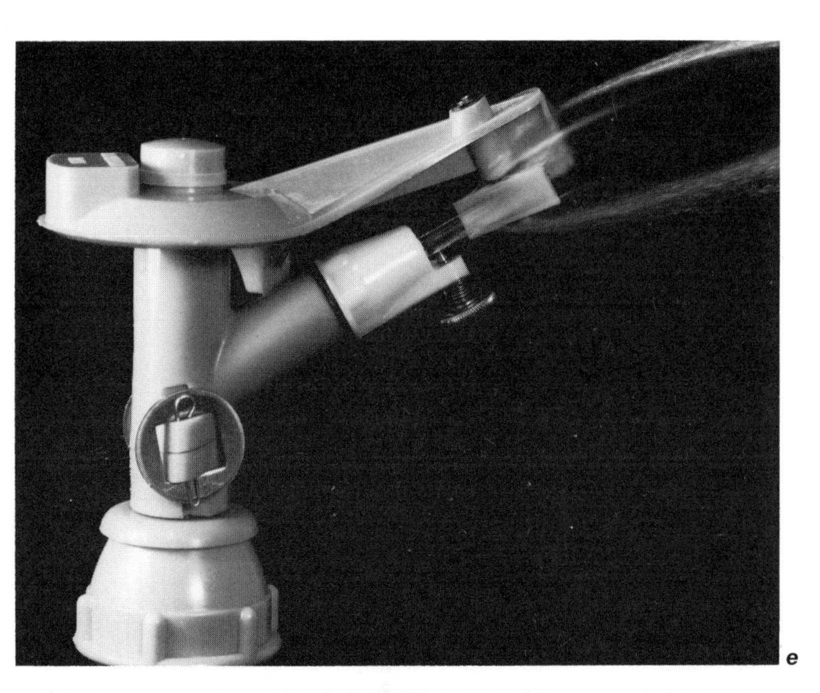

e

e **LAWN SPRINKLER**

Injection-moulded components in nylon. Courtesy Akzo Plastics.

f **SPRINKLER**

Rotating arm is of ABS. The base is of polypropylene. Made by Hozelock Ltd.

g **FOUNTAIN LIGHT**

The lamp body is ABS and the lens unit is polycarbonate. Made by Lotus Water Garden Products.

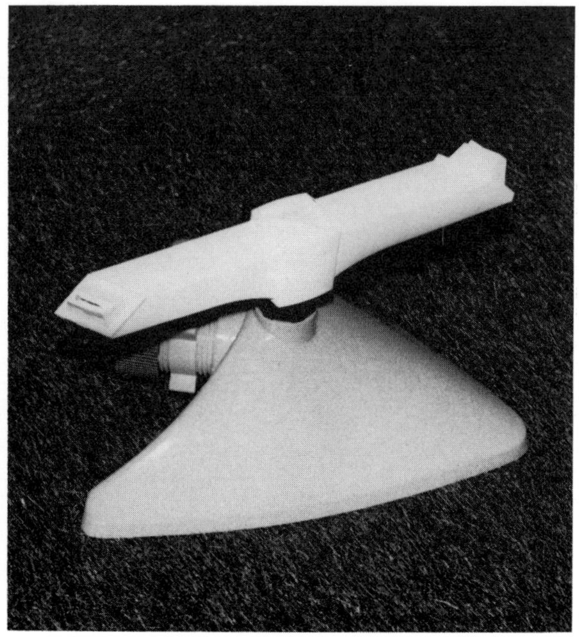

f

g

a **GRASS TRIMMER**

Nylon cord trimmer with a high-impact plastic motor hood. Made by Wolf Tools Ltd.

b **SHRUB TRIMMER**

Operated from a rechargeable battery. Made by Wolf Tools Ltd.

c, e **GRASS CUTTERS**

The hood covers are made in high-impact ABS. The indicator domes are acrylic, both injection-moulded. Made by Flymo Ltd.

b

a

c

d

d **LAWNMOWER**

The grass-box and motor-hood are in high-impact plastics, injection-moulded. Made by Wolf Tools Ltd.

e See previous page.

f **GRASS SHEARS**

One of a range of rechargeable battery-operated tools. Made by Wolf Tools Ltd.

f

e

a **GARDEN POOL**

Vacuum-formed in high-density polythene. Made by Lotus Water Garden Products.

b **GARDEN ITEMS**

All are made by rotational moulding from 'Alkathene', ICI's polyethylene. Courtesy ICI Plastics Division.

a

b

c **TRUG**

Injection-moulded from polythene. Made by Stewart Plastics.

d **WEDGES**

Tree-felling wedges made from nylon. Courtesy Akzo Plastics.

e **RATCHET PRUNER**

The handles are moulded in nylon reinforced with glass fibre. Courtesy CK Tools.

c

d

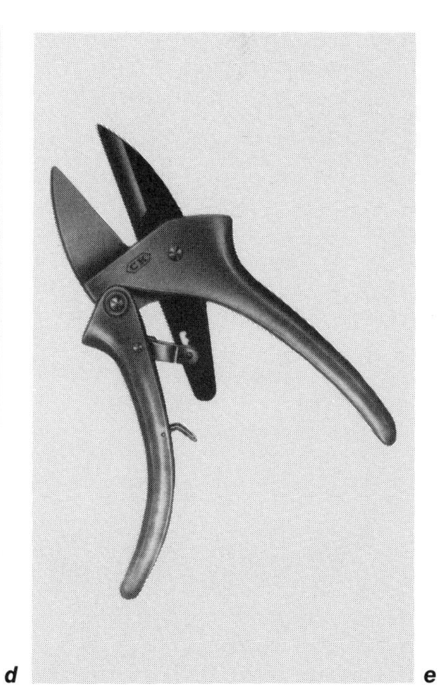

e

a **FOUNTAIN JETS**

These are all moulded in polypropylene by Lotus Water Garden Products.

b, c **WATER FILTER UNIT**

Photo (***b***) shows the unit before the foam pads are added. The unit (***c***) is injection-moulded in ABS and it holds a gravel filtration element. The foam pads are expanded polyurethane, and they remove fine particles from the water. Made by Lotus Water Garden Products.

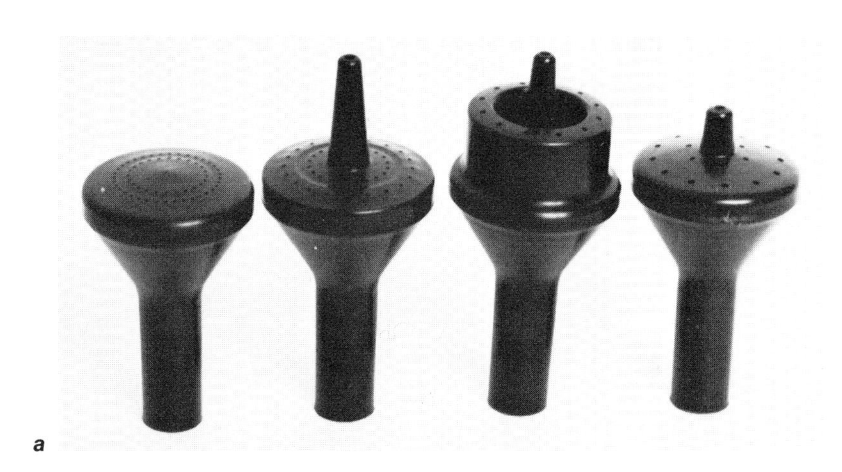

a

b

c

LEISURE, SPORT & HOBBIES

SUN-GLASSES

The frame is made from acetate by O. Goldsmith Ltd. Courtesy Courtaulds Acetate Ltd.

a SAILING YACHT

The hull and deck are of GRP. The sail material is polyester fibre. Made by Westerly Marine Construction Ltd.

b INTERIOR VIEW

The headlining is PVC with a foam backing. Cushions of polyurethane foam. Made by Westerly Marine Construction Ltd.

a

b

c

d

c YACHT WINCH COMPONENTS

Made mainly from plastics. 'Maranyl', ICI's nylon, is used for the majority of the components. Made by Girdlestone Mouldings.

d HOVERCRAFT

A two-man craft designed and produced in GRP by Specialised Mouldings Ltd.

e WIND-SURFER

The shell is moulded in Monsanto's Lustran ABS, which adheres well to the polyurethane foam core. Made by Tabur SA, France.

e

a **SKI-SURFER**

Produced in GRP, the surfer is fitted with a twin-cylinder two-stroke engine plus a water jet unit. Designed and produced by ATM (Ski-Surfer) Ltd.

b **NAVIGATION LIGHTS**

Used on small boats and leisure craft, the case is moulded ABS, the lens is polycarbonate. Made by Lucas Marine.

c **MARINE ACCESSORIES**

These items have been made in 'Kematal' acetal copolymer because of this plastic's resistance to the effects of salt water and temperature change. Made by RWO (Marine Equipment) Ltd.

a

b

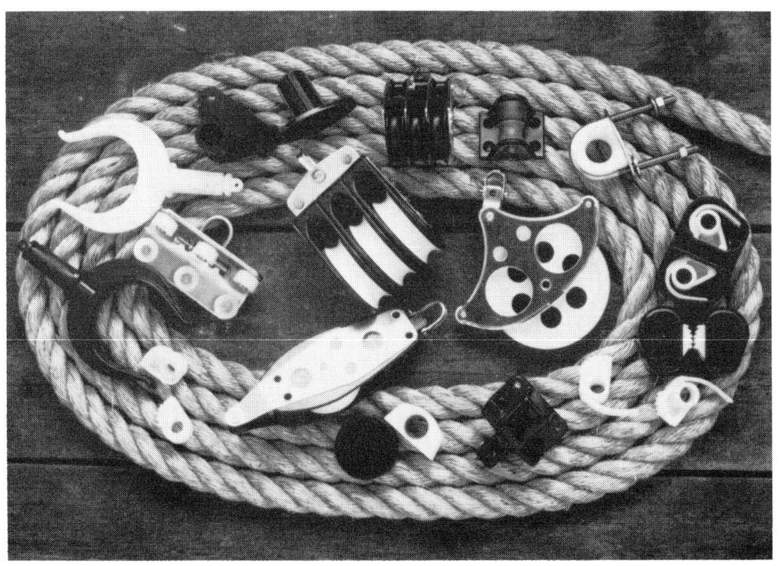

c

d

e

d **FORMULA 1 RACING AIRCRAFT**

Originally a fabric-covered frame, now finished in GRP strengthened with kelvar and carbon fibre, the fuselage and tail unit being redesigned. Moulding work by Specialised Mouldings Ltd.

e **BICYCLE WHEEL**

The wheel is injection-moulded in a glass fibre reinforced grade of nylon. Courtesy ICI Plastics Division.

a

a TENNIS RACKET

The frame is made by injection-moulding nylon reinforced with carbon fibre on to and around a low-melting alloy metal core, this being melted away to give a hollow section which is filled with rigid polyurethane foam. A low density is used in the head to dampen vibration and ensure good balance. Medium density in the handle provides a rigid base for the grip. Epoxy resin is used to give a tough durable finish to the racket. Courtesy Dunlop Ltd.

b RACKET FRAME

The alloy frame is melted out to leave a hollow frame with internal support columns and reinforced stringing holes around the head. Courtesy Dunlop Ltd.

c BADMINTON RACKET

The frame is made of four plies of beech and ash with two plies of carbon fibre. Made by Stiga A.B. Sweden. Courtesy Carbon Fibres Division, Courtaulds Acetate Ltd.

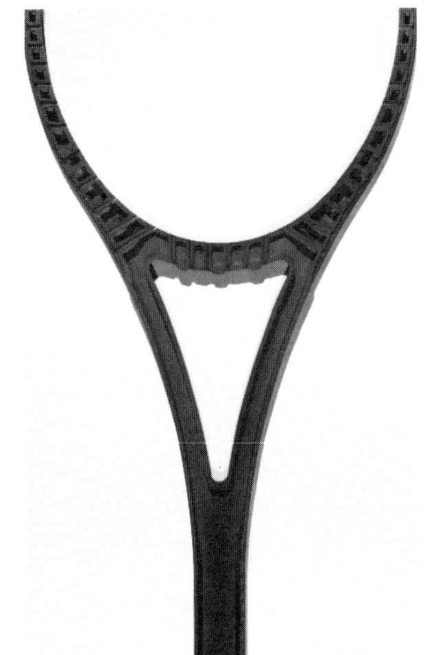

b

c

d

e

d GOLF BALL

High-speed photograph showing the degree to which the ball is deformed at the moment of impact. Courtesy Royal Military College of Science.

e GOLF CART

The wheel tyres are rotationally moulded in uPVC. The wheel centres being injection-moulded in PVC or nylon. Made by Dawman Ltd.

f GOLF CLUBS

The ferrules of these clubs have cellulose acetate butoryl in their construction. Made by Dunlop Sports Co. Ltd.

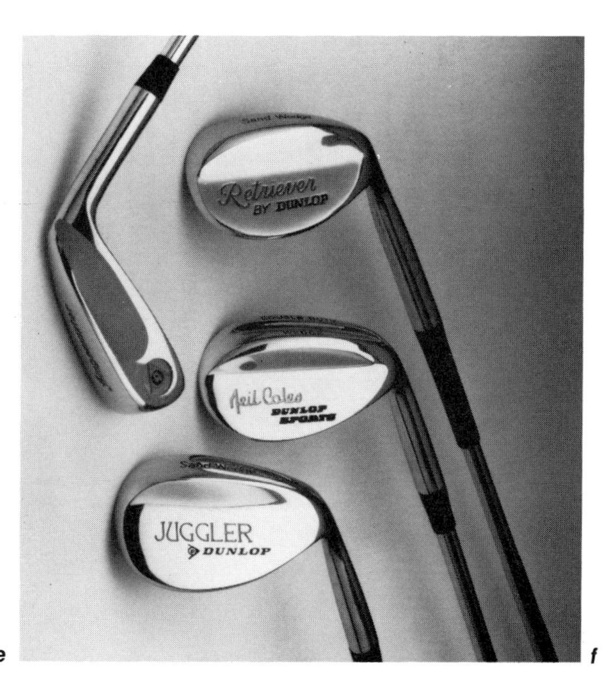

f

a SKIS: DOUBLE BOX CONSTRUCTION

This type of ski has an acrylic foam core reinforced with carbon fibre and fibreglass, and a polyethylene sole. The diagram (inset) shows the method of construction. Made by Madshus Skifabrikk A/S, Norway. Courtesy Courtaulds Acetate Ltd.

b SKIS: MULTI-TORSION BOX CONSTRUCTION

Lengths of lightweight timber are separated by fibreglass ribs and strengthened by carbon fibre rods, assembled as shown in the diagram below. The ski has an outer casing of epoxy resin and a polyethylene sole. Made by Vielhaber Skis Ltd.

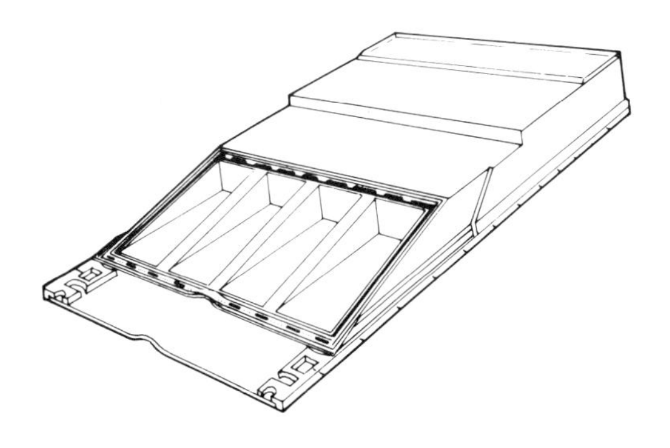

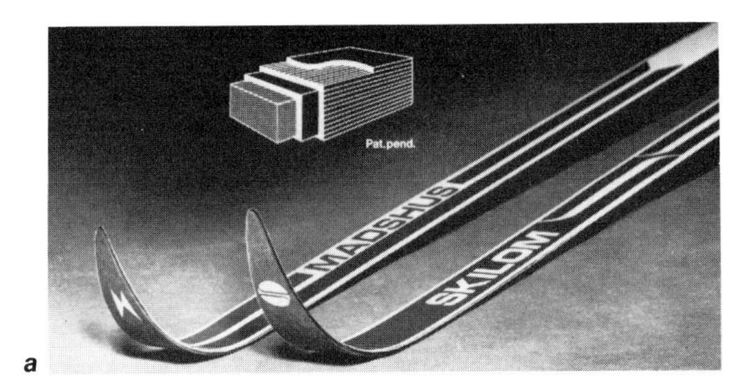

a

b

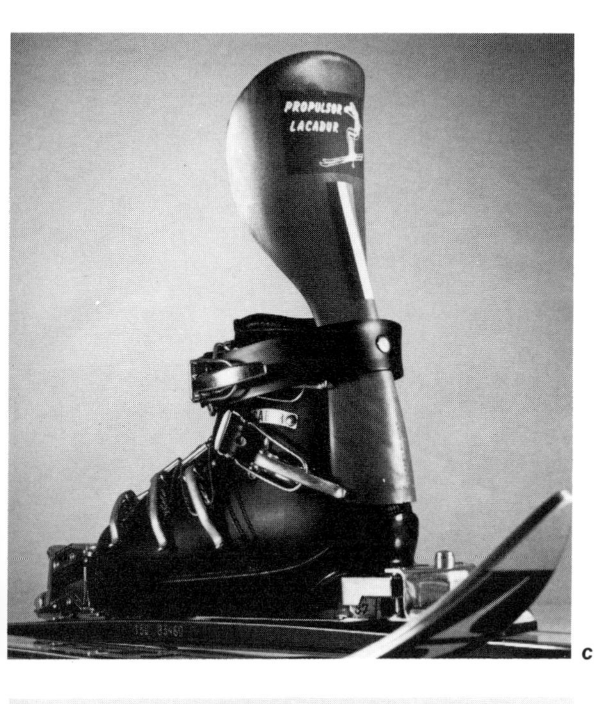

c

c **SKI-BOOT**

The heel-piece is moulded in 'Arnite' Akzo's thermoplastic polyester. Courtesy Akzo Plastics.

d **FOOTBALL BOOTS**

The soles and studs are one moulding of PVC. Made by Clarks Ltd.

e **WELLINGTONS**

The boots are injection-moulded in PVC. The drawstring tops are a PVC-coated fabric. Courtesy Dunlop Ltd.

d

e

a **SOFT-SIDED CASE**

Made of scuff-resistant vinyl by Custom Synthetics Ltd.

b, *c* **SHOES**

The 'nature trek' and 'polyveldt' shoes have expanded polyurethane soles, which are produced by rotary injection-moulding. Made by Clarks Ltd.

d **LADIES' SHOES**

The soles are moulded in PVC. Made by Clarks Ltd.

a

b

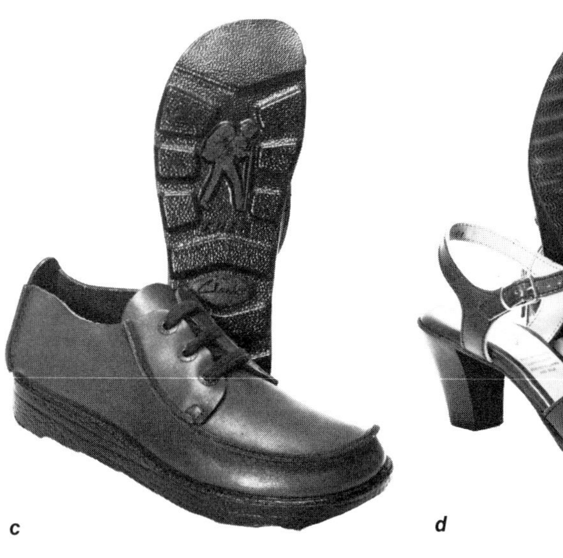

c

d

e

e EXECUTIVE CASE

Moulded in ABS. Made by Custom Synthetics Ltd.

f CITY CASE

Made in 'leather look' expanded vinyl and aluminium by Custom Synthetics Ltd.

g INSTRUMENT CASE

Moulded from ABS and fitted with an aluminium gasket rim. The interior foam is cut to suit the equipment being carried. Made by Custom Synthetics Ltd.

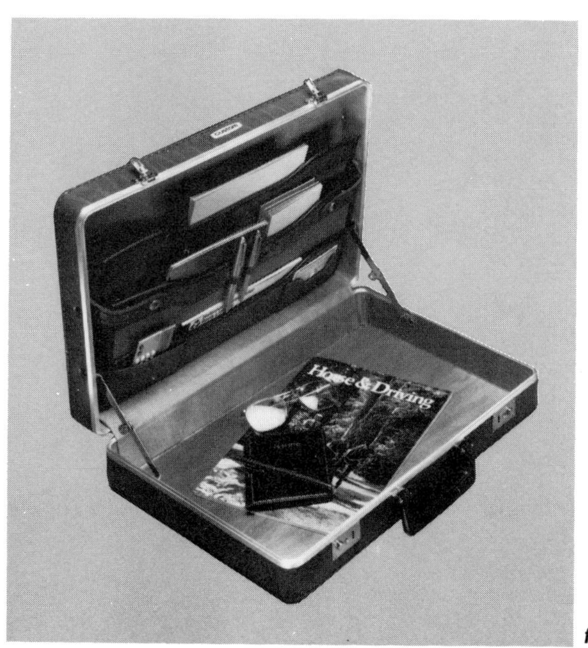

f

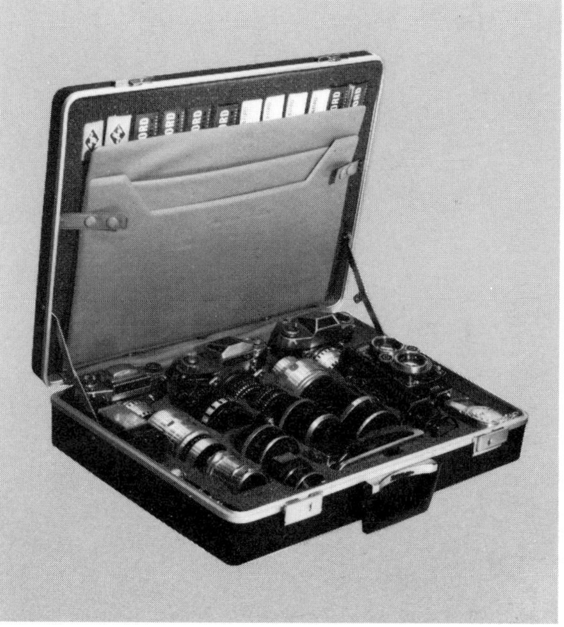

g

a RECORD AND CASSETTE RACKS

These racks are moulded in ABS. Made by Crayonne.

b CLARINET

The body is moulded entirely in fibreglass reinforced nylon. Courtesy Fibreglass Ltd.

c LEISURE ITEMS

All made from various plastics. Courtesy ICI Plastics Division.

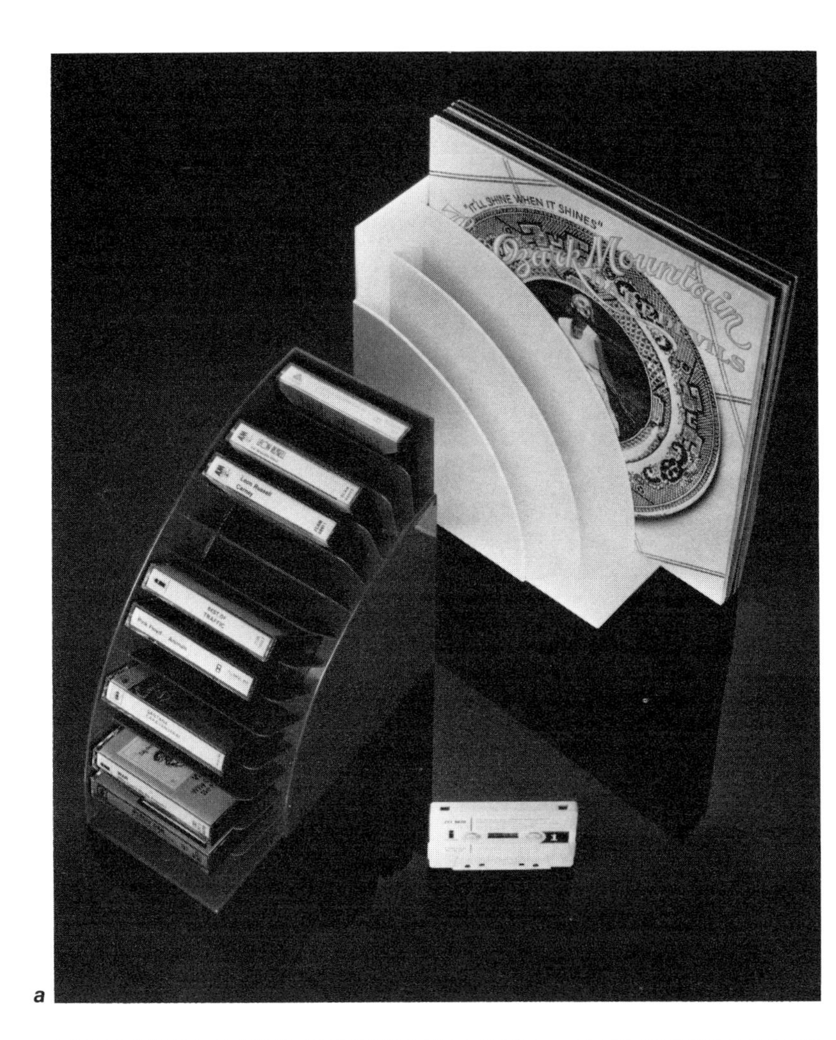

a

b

c

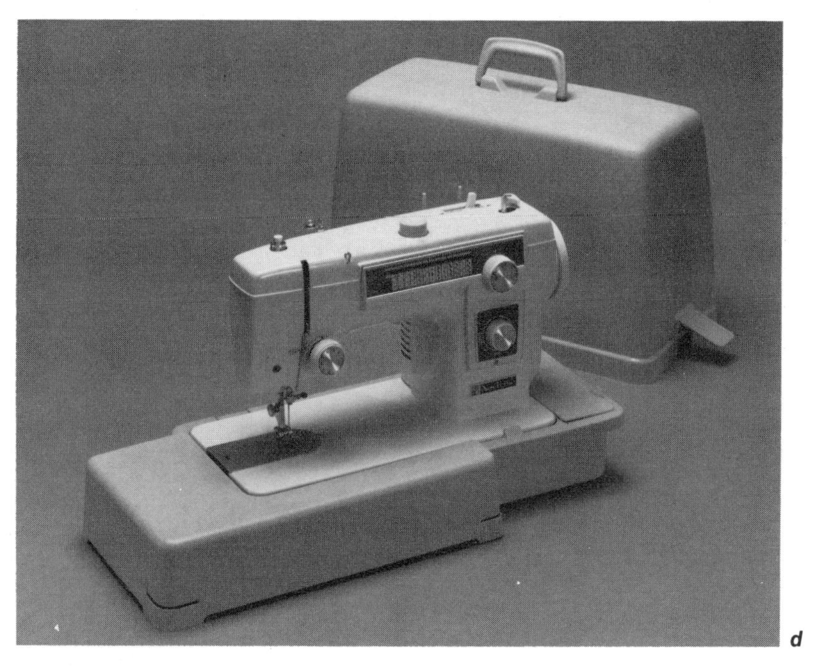
d

d **SEWING MACHINE**

Moulded in polypropylene. Integral hinges are incorporated into the base moulding to locate the machine and in the catches for the carrying case. Made by WCB Plastics Ltd. Courtesy ICI Plastics Division.

e **STAMP GUILLOTINE**

This miniature guillotine is moulded from Monsanto's ABS. Courtesy Monsanto Ltd.

f **HOBBY KEEPER**

A general-purpose storage box. Made by The Tupperware Co. Ltd.

e

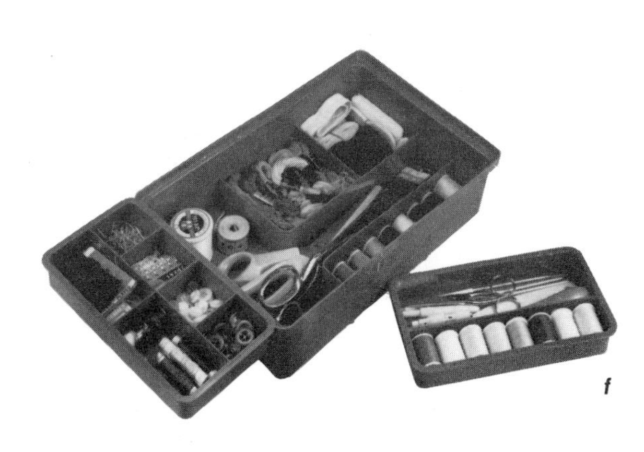
f

MODERN DESIGN IN PLASTICS

a TOOL CADDY

Injection-moulded in polythene. Made by Stewart Plastics.

b HOBBY SNIPS

For general-purpose use. The handle is polypropylene. Courtesy CK Tools.

c TRIMMER'S SCISSORS

The polystyrene handle is injection-moulded around the blade. Courtesy CK Tools.

d RASP

Handle of polyethylene, into which the blade push fits. Courtesy CK Tools.

e TAPE MEASURE

The two halves of the shell are moulded in ABS and metal plated. The windows are clear polycarbonate. Courtesy CK Tools.

f TOOLBOX

Made in ABS, injection-moulded. Courtesy CK Tools.

a

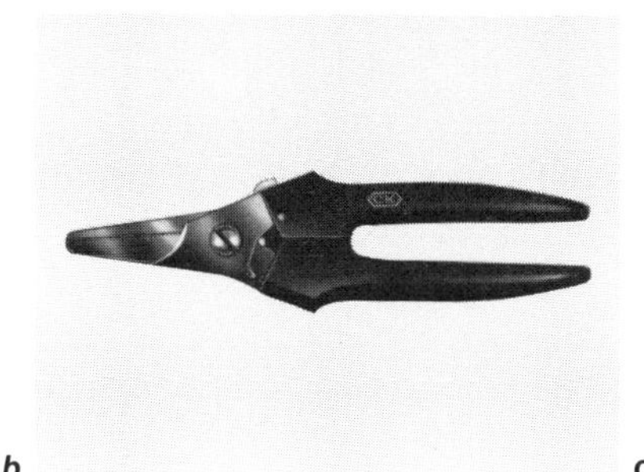

b

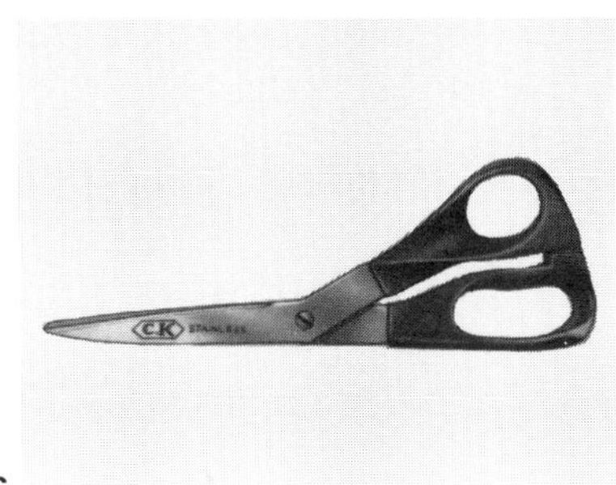

c

f

d

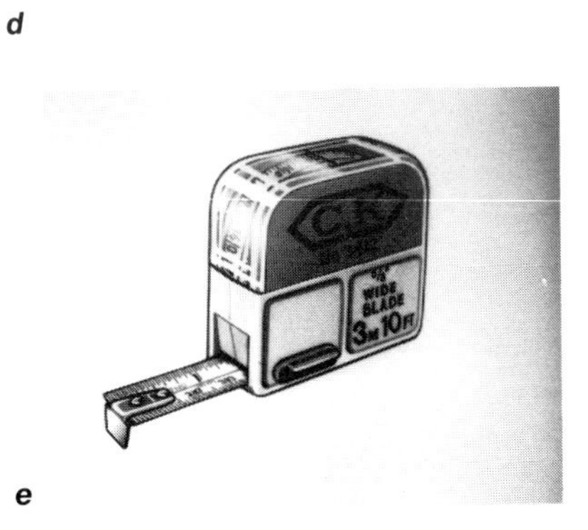

e

NURSERY PRODUCTS & TOYS

ROCKING HORSE

A rigid plastic base is upholstered in 'Repol', Dunlop's chipfoam, then topped with a layer of 'Dunlopreme' polyether foam. Courtesy Dunlop Ltd.

a **BIB**

Moulded in polypropylene. Made by Cindico.

b **CAR SEAT**

The seat is injection-moulded in polypropylene. Made by Cindico.

c **SUPER SEAT AND STAND**

The seat is moulded in polypropylene. The knobs and cones are of nylon. Made by Cindico.

a

b

c

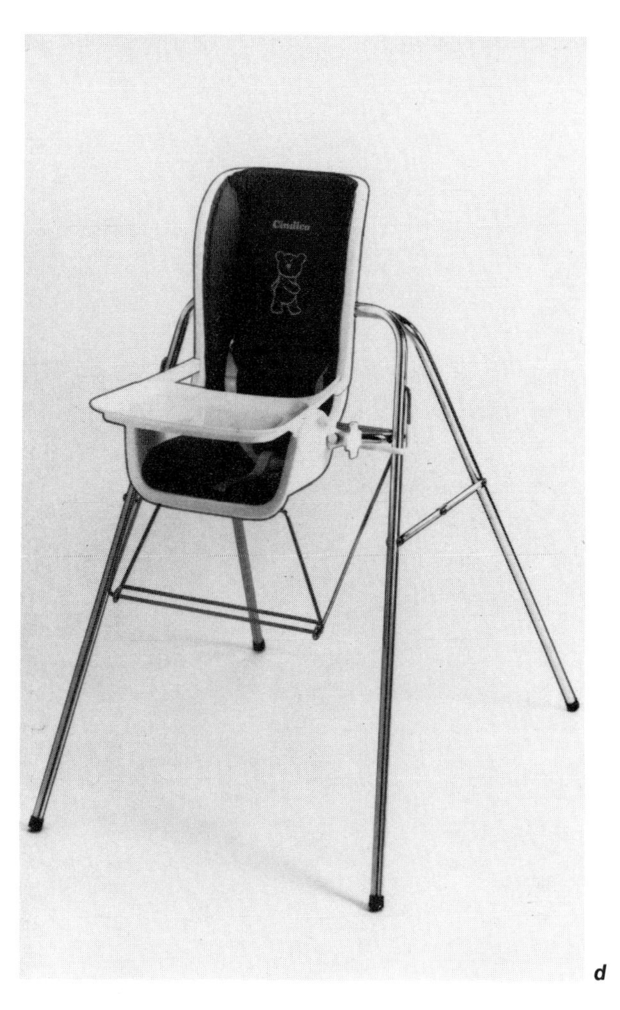

d

d CADET SEAT

Seat and tray injection-moulded from polypropylene. The knobs and cones from nylon. Made by Cindico.

e, f DOUBLE LI-BAK FOLDING CHAIR

The handgrips are dip-moulded PVC. The seat pad and footrest mouldings are from polypropylene. The spacers, sliding-locking mechanism and armrests are nylon, the armrest badge being polypropylene. The struts are glass-reinforced nylon. Made by Cindico.

e

f

a **CINDIPIN**

Injection-moulded in polypropylene. Made by Cindico.

b **FLIP FINGERS**

The handle is ABS, the fingers are polypropylene and the axle is nylon. All injection-moulded. Made by Hestair Kiddicraft Ltd.

c **BANGLE RATTLE**

The rings are ABS and charms polythene. Made by Hestair Kiddicraft Ltd.

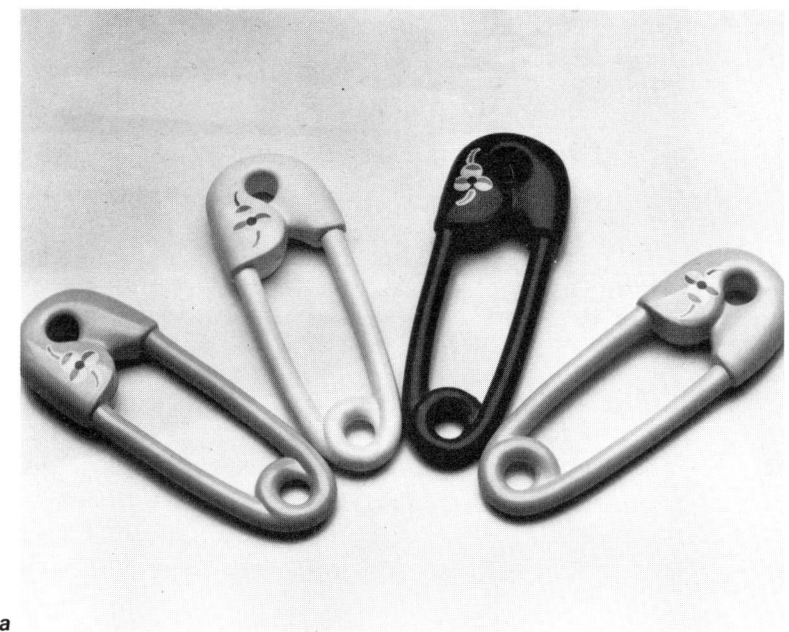

a

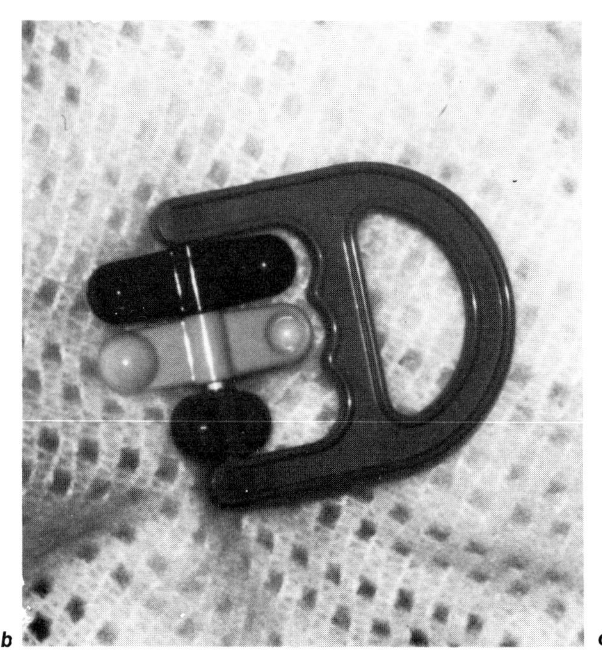

b

c

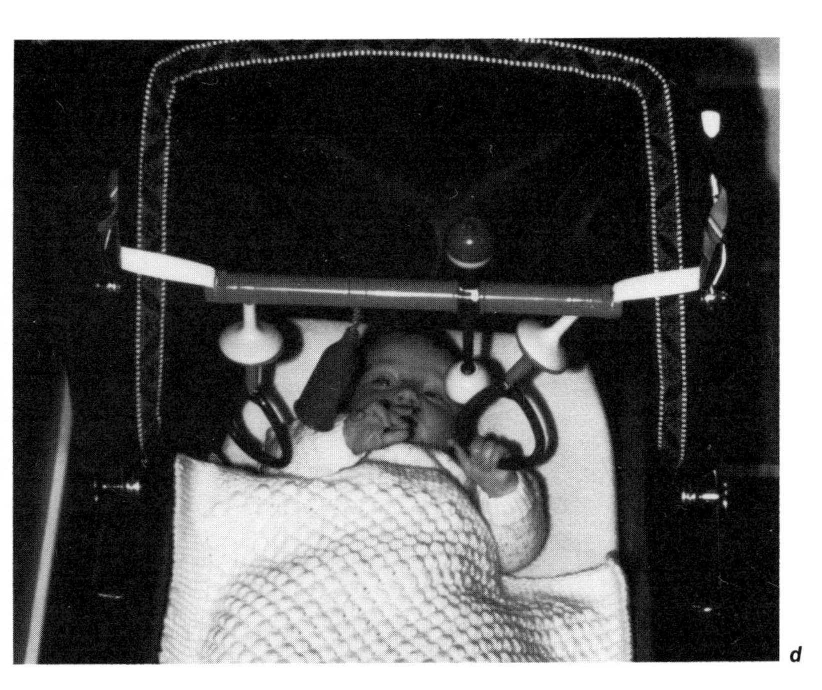

d

***d* PRAM/CRADLE PLAY**

Injection-moulded, the various components are of ABS, polypropylene and high-impact styrene. Made by Hestair Kiddicraft Ltd.

***e* CRAWL-A-BALL**

This soft-textured ball is rotationally cast in PVC. Made by Hestair Kiddicraft Ltd.

***f* RATTLE**

The ball and bodywork are made from polypropylene. The collar is ABS. All injection-moulded. Made by Hestair Kiddicraft Ltd.

e

f

a **BABIES FEEDING-BOTTLE TEATS**

Moulded in 'Evatone', ICI's EVA resin, by Colgate Plastics.

b **PAINT POTS**

These non-spill pots with their tapered stoppers are made from polypropylene. Courtesy E J Arnold & Son Ltd.

a

b

c

c KITCHEN TOYS

The bowler-hatted figure has feet that act as biscuit-cutters, while the body holds measuring spoons and the hat itself is a flour-shaker. The arms of the figure on the left act as a pair of scales, while the head is a mixing-bowl. All made in polypropylene by Peter Pan Playthings.

d LEARNING TO SPELL

The letters are magnetic and store away into the main frame which is of high-impact styrene. Made by Peter Pan Playthings.

e LEARNING CLOCK

The frame is moulded in high-impact styrene. As the minute hand passes over each hour the numbers automatically flip over to reveal the minutes. Made by Peter Pan Playthings.

d

e

a **TALLY BOARD**

Made of polystyrene, the tray being vacuum-formed. Courtesy E J Arnold & Son Ltd.

b **STRATEGY GAMES**

The games boards are moulded from high-impact styrene. Made by Peter Pan Playthings.

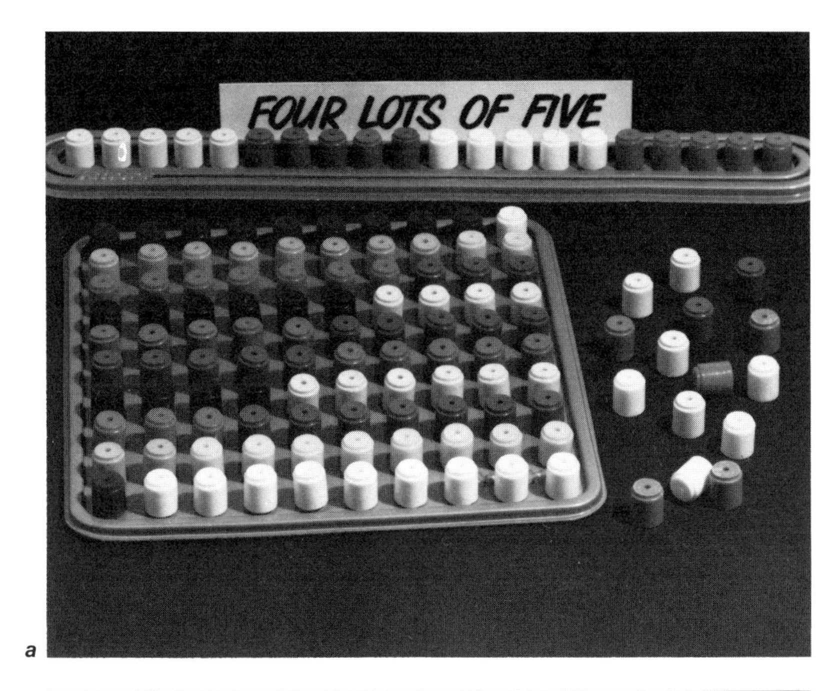

a

b

c

c MODEL LOCOMOTIVE

Injection moulded in polystyrene. Made by Hornby Hobbies.

d MULTI-TERRAIN VEHICLE

Made from high-impact polystyrene, injection moulded by Palitoy Ltd.

e TURBO COPTER

Made by Palitoy Ltd, in high-impact polystyrene.

d

e

a **MASTER MIND**

Pegs are polypropylene, board and shield from high-impact styrene. Made by Invicta Plastics Ltd.

b **ELECTRONIC MASTER MIND**

Carried in a flexible PVC wallet. Made by Invicta Plastics Ltd.

a

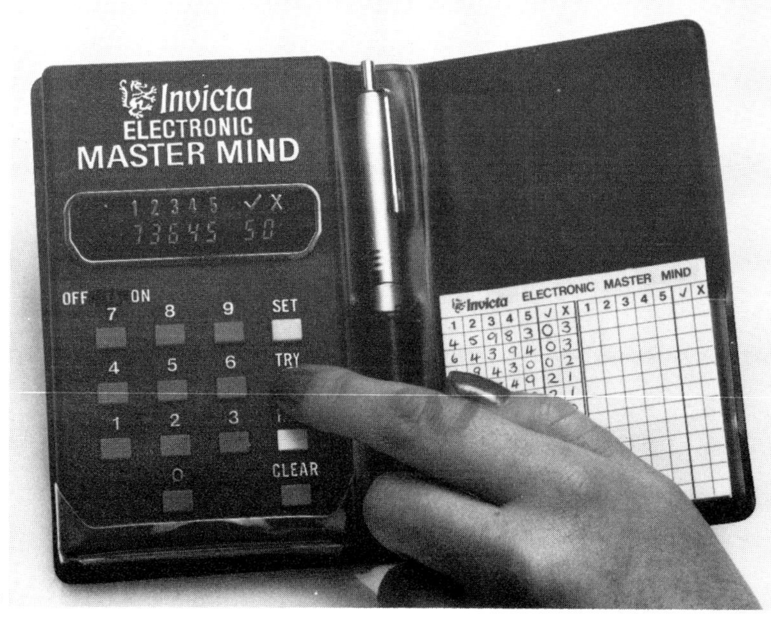

b

FURNITURE & INTERIORS

EASY CHAIR DOLL

The chair interior is filled with crumbed 'Dunlopreme' polyether urethane foam. The linking rings can hold the doll in several different positions. Courtesy Dunlop Ltd.

a **CHAIR**

The first moulded polypropylene chair seat. Designed by Robin Day for Hille in 1964, and still in production.

a

b **BED**

Upholstered headboard and a peripheral seat fitted to the sides and foot of the bed. Photo by ICI Plastics Division.

c **BED**

Ten structural foam mouldings of 'Propathene', ICI's polypropylene, interlock to form a rigid structure including the headboard and side-tables. Made by Cabinet Industries Ltd. Photo by ICI Plastics Division.

b

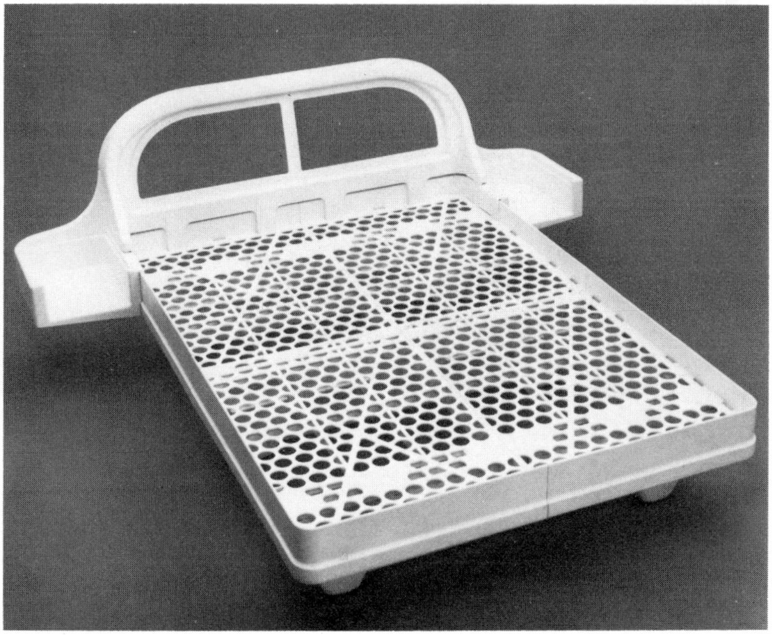

c

a **CHAIRS**

Injection-moulded polypropylene chairseats with upholstered seat and back pads, these having a foam interior covered with expanded vinyl or wool fabric. Metal framework finished in epoxy, nylon or chromium-plated. Made by Pel Ltd.

a

b

c

***b*, *c* PATIO FURNITURE**

The range of furniture is made from junctions and piping of ABS. The table top is coated with melamine. Made by Rainbow Associates.

a, ***b***, ***c*** **CHAIRS**

The seat shells are formed in GRP. Flexible polyurethane foam is then moulded around the shell to give body contour. Various underframes can then be attached, these being of chromed steel or aluminium. Made by Yorkshire Fibreglass Ltd.

a

b

c

d

d **UNDERSIDE OF SEAT**

The special design gives a flexible stability.

e **GENERAL-PURPOSE CHAIR**

This chair may be stacked for storage and interlocked for mass seating arrangements. The seat and back is injection-moulded in polypropylene. The wooden legs are beech. Made by Carl Sasse GmbH & Co, West Germany.

f **CONSTRUCTION OF CHAIR**

Simplicity: there are no screws or fittings required.

e

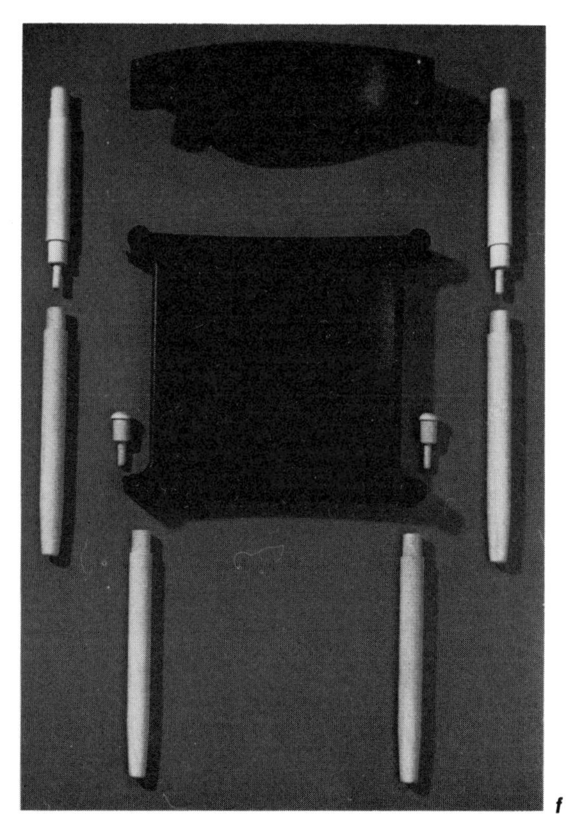

f

a **TOP OF PLASTIGRAPH TROLLEY**

b **THE PLASTIGRAPH TROLLEY**

Designed for multi-purpose use, this and the Boby are made in toughened styrene by Bieffeplast, Italy.

c **THE BOBY TROLLEY**

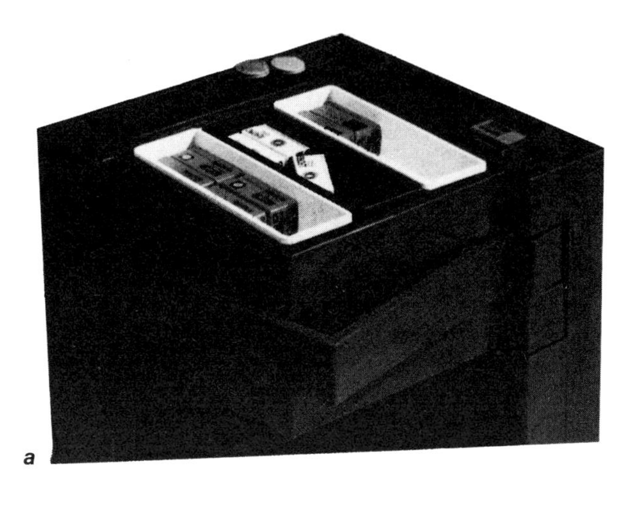

a

b

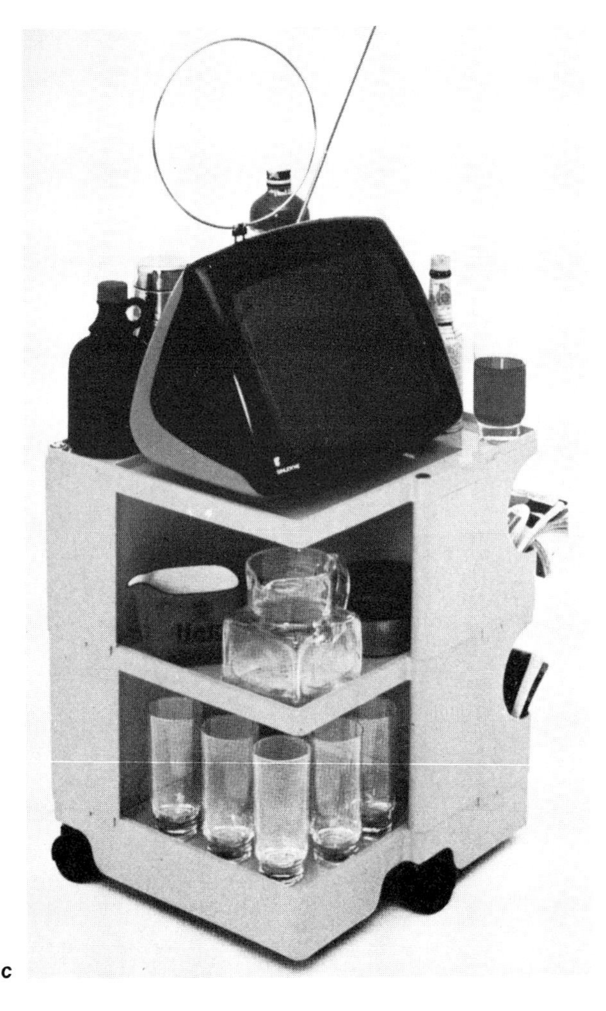

c

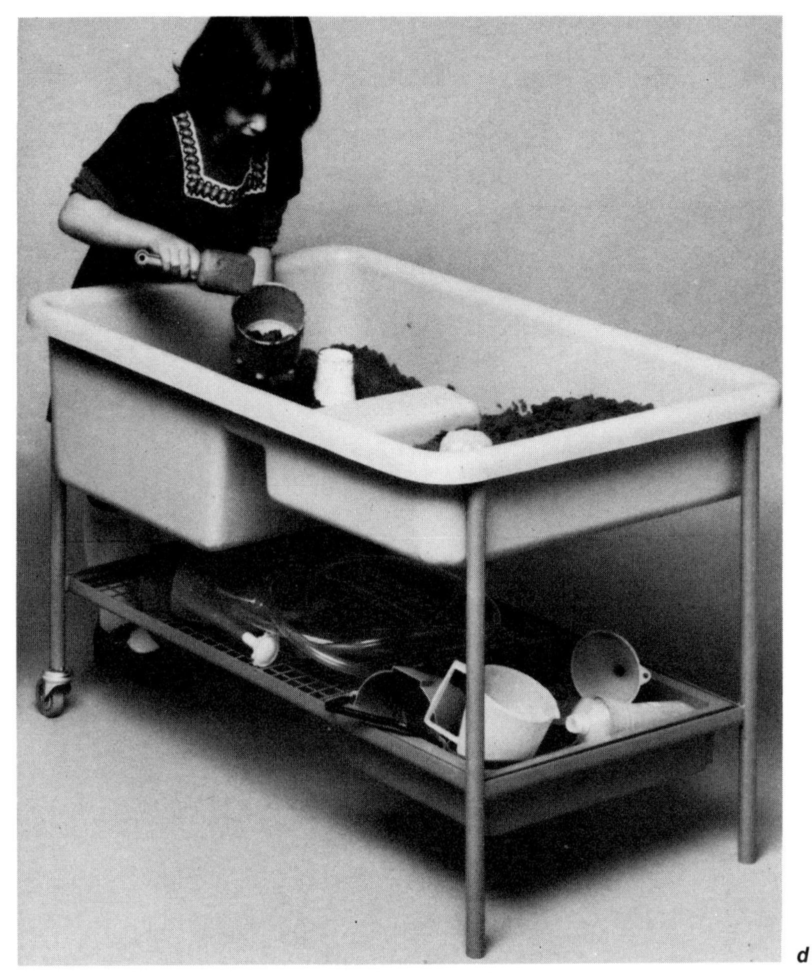

d

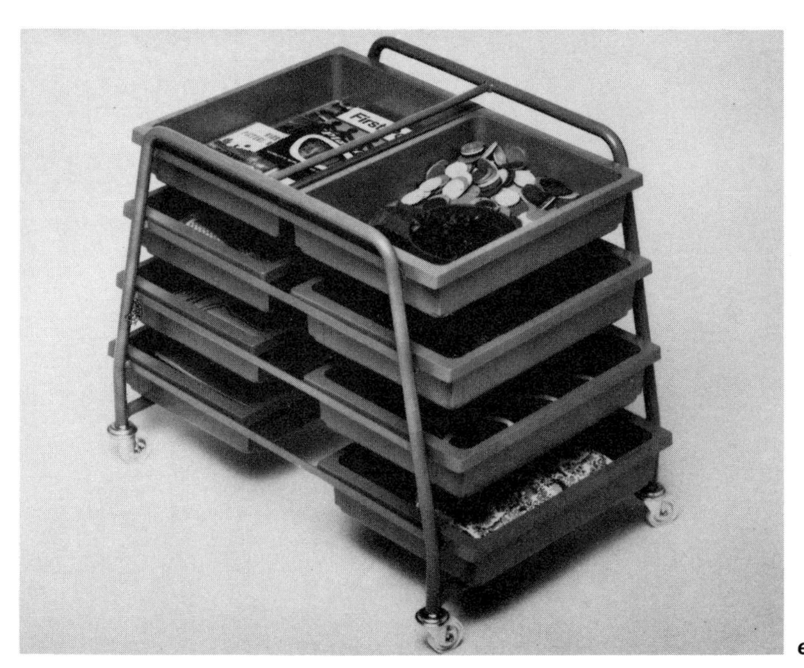

e

d SAND/WATER TROLLEY

Polyethylene is used for the top container. The frame is epoxy coated and the base storage tray is of high-impact styrene. Made by Pel Ltd.

e TRAY CART

The trays are moulded in high-impact styrene. Steel tubular framework is coated in epoxy resin. Made by Pel Ltd.

a **OFFICE CHAIR**

Arms are cast aluminium coated with nylon. The seat is high-impact styrene. Made by Verco Office Furniture. Courtesy Matthews Office Furniture.

b **ACRYLIC FURNISHINGS**

The wall in the background comprises three distinct layers of hand-painted clear 'Perspex', ICI's acrylic sheet. The table lamps (including their shades), the standard lamp and the chandelier are made entirely from either clear or opaque coloured perspex. Courtesy ICI Plastics Division.

a

b

c

c SEAT

Prototype design with a laminated ash frame. The seat and back are cylinders of foam using various grades of polyether urethane foam. Courtesy Dunlop Ltd.

d FURNISHING MATERIALS

The carpets, settee, table, mirror, light fittings, the television cabinet veneer and the garden furniture are all made from a variety of plastics. Courtesy ICI Plastics Division.

d

***a*, *b*, *c* CABINET UNITS**

The rolling shutter door of these cabinets are made of ABS. Marketed by Banks Heeley.

***d* STORAGE CABINET**

The cabinet and sliding tambour is made from acrylic sheet. Made by Banks Heeley.

a

b

c

d

e

f

e, f KITCHEN UNITS

The working tops and fascia material of these cabinets are melamine. Made by Crosby Kitchens Ltd.

a **BED/SOFA**

This convertible's interior is of polyether foam. Courtesy Dunlop Ltd.

b, c **FLOORCOVERINGS**

The floor areas are covered with tile squares made from vinyl with a self-adhesive backing. Courtesy Dunlop Ltd.

a

b

c

HOUSEHOLD ITEMS

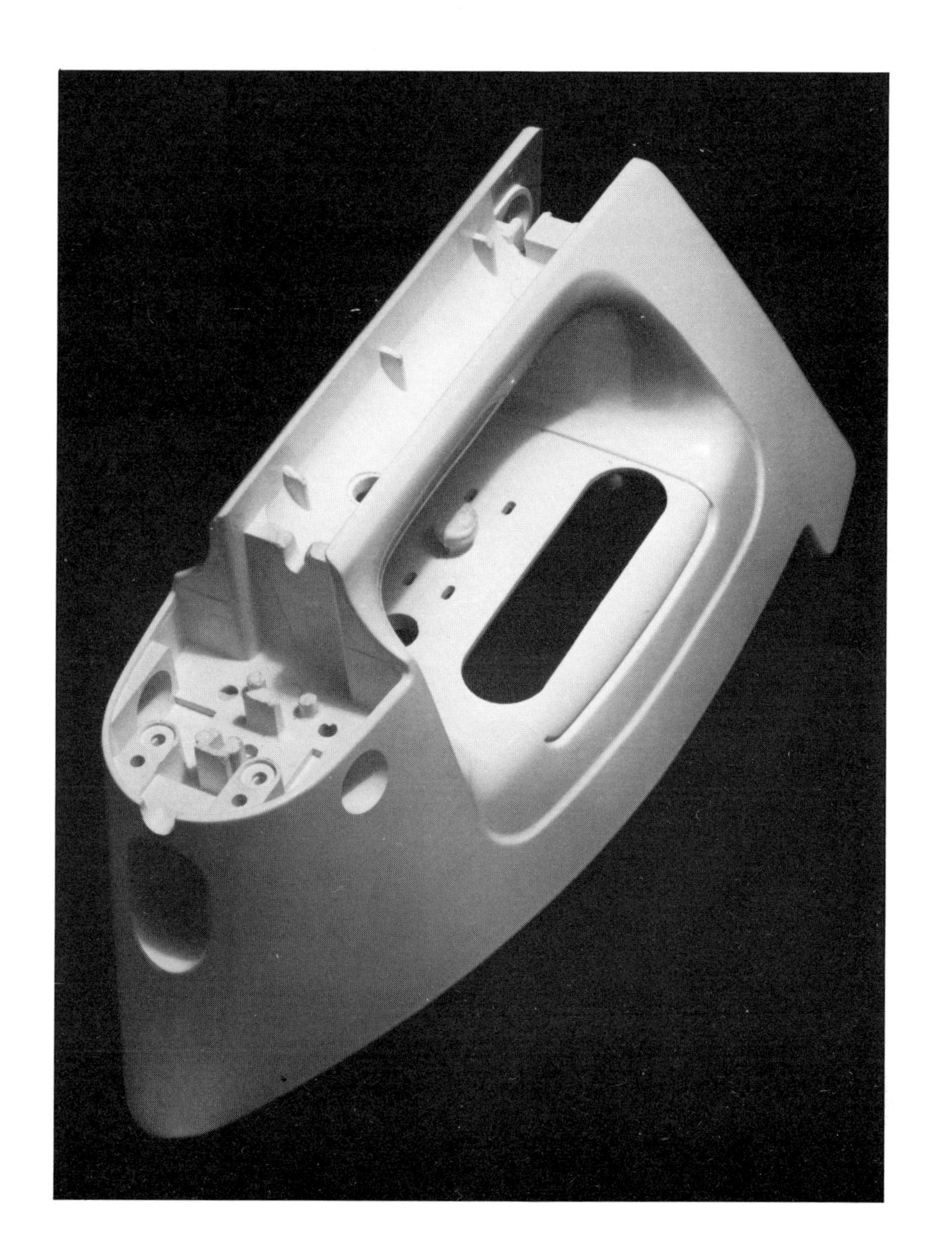

STEAM IRON

The body is moulded entirely in GRP. Courtesy Fibreglass Ltd.

***a* STEAM IRON**

The removable transparent water-tank is made from two pieces welded together. Produced in 'Makrolon', Bayer's polycarbonate. Courtesy Bayer UK Ltd.

***b* TOASTER**

The control dial and moulded end-plates are in phenolic resin. Made by Thorn Domestic Appliances.

***c* TOASTER**

The base is of thermoplastic polyester with a vitreous enamelled-steel casing. Made by Russell Hobbs Ltd.

a

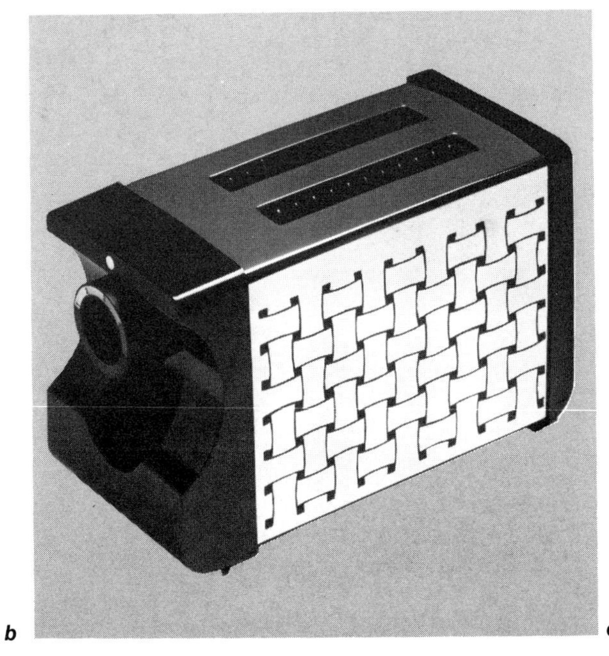

b

c

d ELECTRIC KETTLE

The body is manufactured from acetal copolymer. Made by Russell Hobbs Ltd.

e ELECTRIC KETTLE

Lid and handle of this Hoover kettle are moulded with a textured finish in 'Kematal' acetal copolymer. Courtesy Amcel Ltd.

f ELECTRIC KETTLE

Moulded body is made from GEC's 'Noryl'. Made by Redring Electric Ltd.

d

e

f

a **HAND WHISK**

Battery operated, the body is of ABS. The beater blades are acetal. Made by Thorn Domestic Appliances.

b **GRILL**

The handles, integral with the body are moulded in thermoplastic polyester. Courtesy Akzo Plastics.

c **COOKPOT**

The body, base and lid are moulded from polycarbonate. The lid knob is ABS. The cooking pot is stoneware with a wide carrying rim. Made by Thorn Domestic Appliances.

d **FOOD PROCESSOR**

The casing, bowl and cover are moulded in polycarbonate, the spatula being polypropylene. Made by Pifco Ltd.

e **FOOD PROCESSOR**

The bowl and lid are smoked acrylic with a moulded body of ABS. Made by Thorn Domestic Appliances.

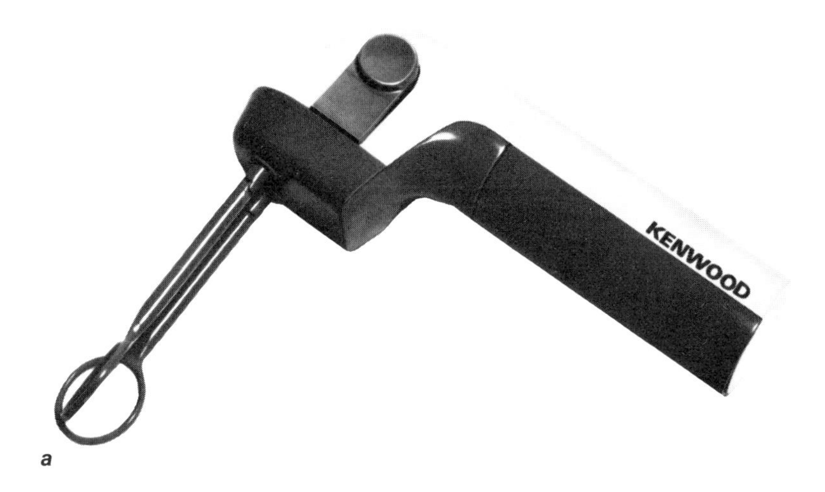

a

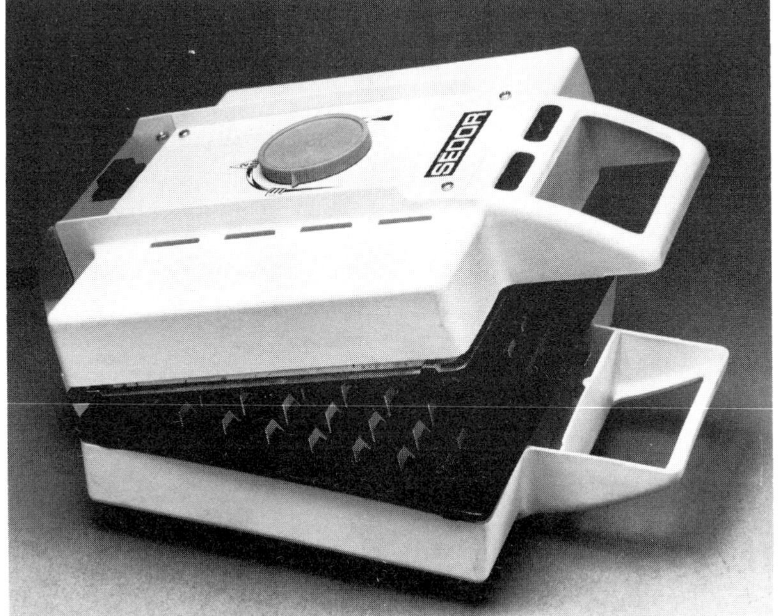

b

c

d

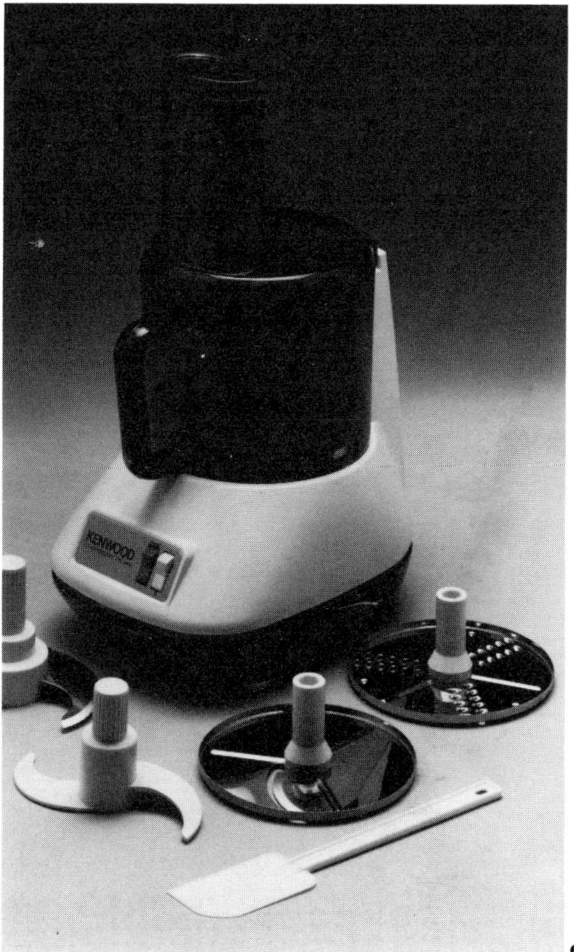

e

a **HAMBURGER PRESS**

Made from polyethylene. Courtesy of The Tupperware Co.

b **KITCHENWARE**

These products are produced in ABS for Marks and Spencer. Courtesy Monsanto Ltd.

a

b

c

d

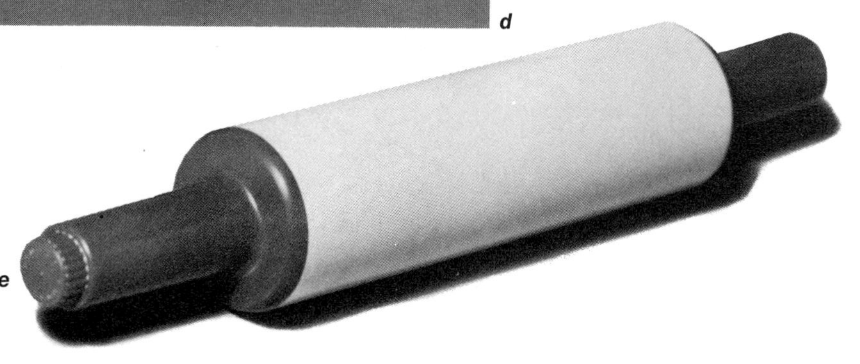

e

c TABLE ACCESSORIES

The oval tray, ashtray and ice-bucket are in ABS. Made by Crayonne.

d PEPPER AND SALT MILLS

Injection-moulded in acrylic. Made by Cole & Mason Ltd.

e ROLLING PIN

Injection-moulded in polystyrene. Made by Stewart Plastics.

a **STORAGE JUGS**

Moulded in polypropylene, the jug having a hinged cover over the spout. Made exclusively for Woolworths by Airfix Plastics.

b **KITCHENWARE**

The salad server, serving tongs and the stacking cruet containers are in polyethylene. Courtesy The Tupperware Co.

a

b

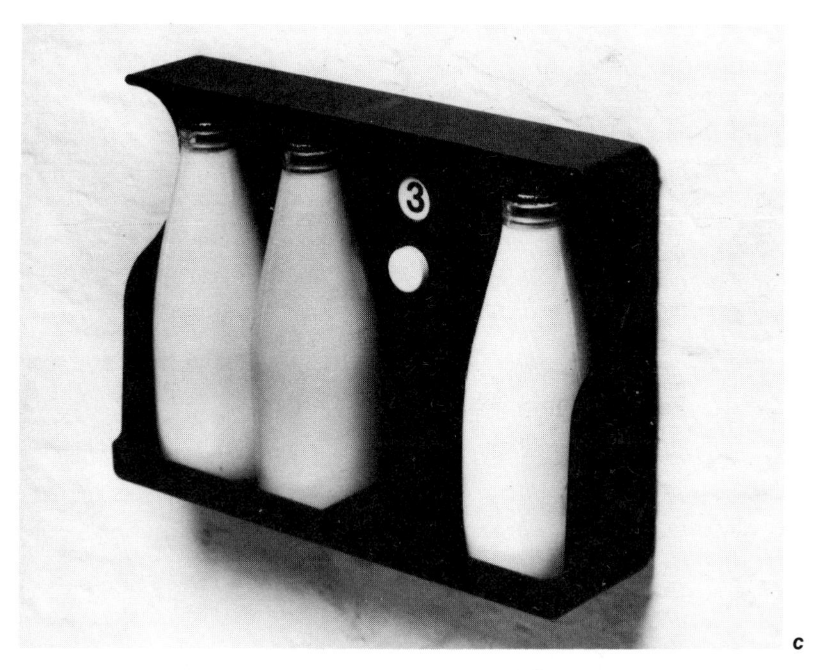

c

c MILK BOTTLE HOLDER

Moulded in polystyrene with a number indicator for the bottles required. Made by Stewart Plastics.

d JUG

Moulded in ABS. Made by Stewart Plastics.

e FOOD MIXING BOWL

Moulded in acetal copolymer for the Kenwood Chef mixer. Courtesy Amcel Ltd.

d

e

a LIGHT/CLOCK

The case is injection-moulded in ABS by Metamec Ltd.

b WALL CLOCK

Moulded in plating grade ABS, by Metamec Ltd.

c DIGITAL CLOCK

The case is moulded in ABS by Metamec Ltd.

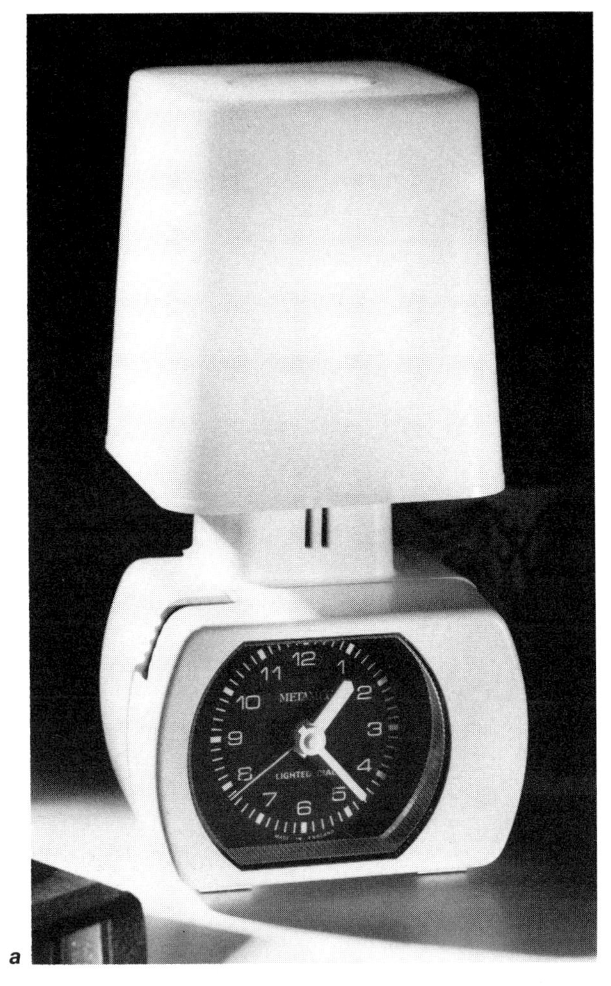

a

b

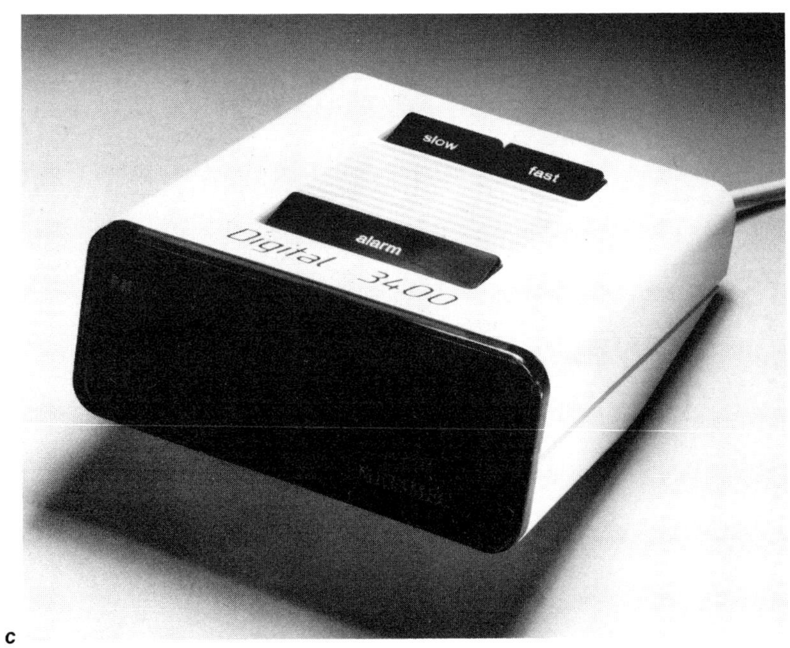

c

d **CLOCKS**

Commissioned by Christopher Strangeways, designed by Keith Gibbons. Photo courtesy Design Council.

d

a **DESK LAMP**

Moulded for Hanimex in a high-gloss grade of ABS. Courtesy Monsanto Ltd.

b, c **HANDLAMPS**

These are both injection-moulded in ABS. Made by Ever Ready.

a

b

c

d

_d_, _e_, _f_, _g_ DOOR CHIMES

The fascia covers are all injection-moulded using toughened polystyrene. The 'Seville' has a bright orange disc on a matt black cover. 'Apollo' has a choice of three coloured centre inserts on a white cover. All of the styles may be mains or battery operated. Made by Friedland Ltd.

_d_ SEVILLE

_e_ APOLLO

_f_ ECHO

_g_ DUET ELECTRONIC CHIME

e

f

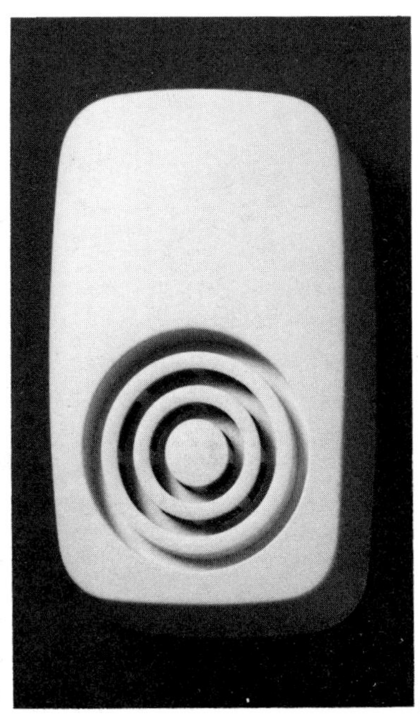

g

a **ASTROFON TELEPHONE**

Electronic push-button phone made from ABS. Courtesy Thorn-Ericsson Telecommunications.

b **DEEP HEAT MASSAGER**

Casing and heater pad are in ABS. The applicators are PVC and polyethylene. Made by Pifco Ltd.

c **ERICOFON TELEPHONE**

The world's first one-piece telephone produced in ABS. Made by L.M. Ericsson, Sweden.

a

b

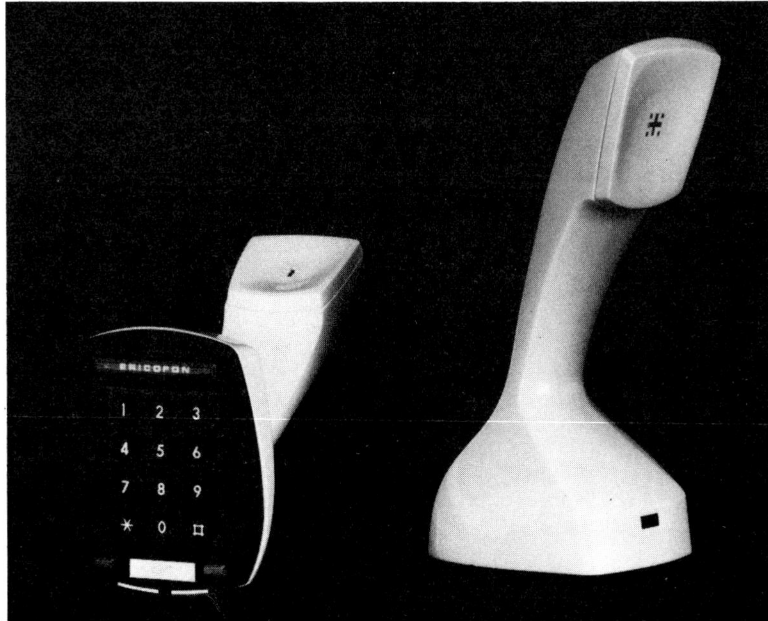

c

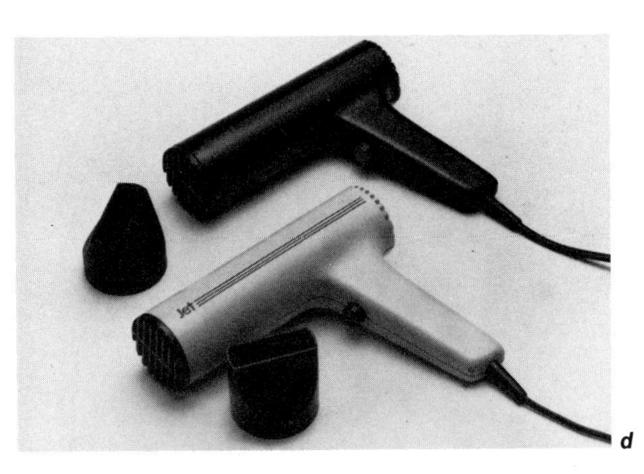

d

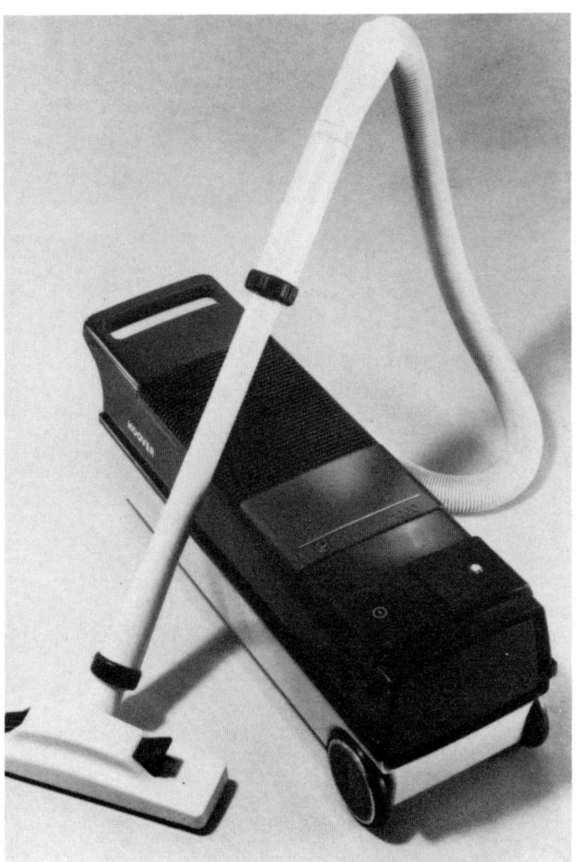

e

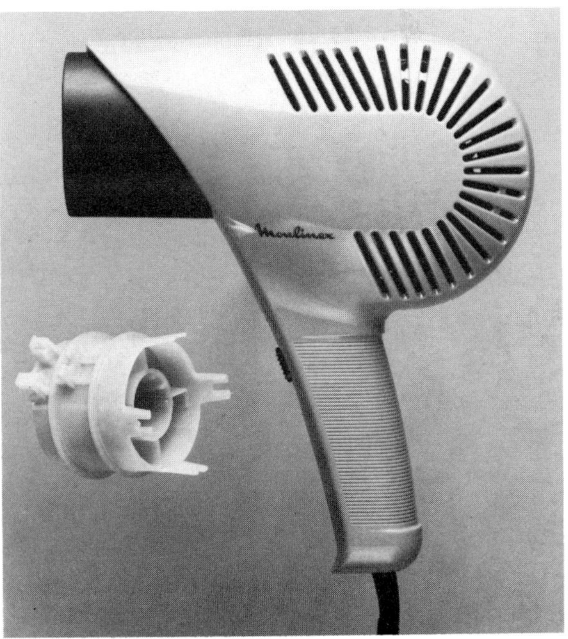

f

d **HAIRDRYERS**

The casings are ABS, the outlet grilles and blow waves are glass-filled nylon. The cord strainer is polyethylene. Made by Pifco Ltd.

e **VACUUM CLEANER**

The body casing, head and tubes are all moulded in Monsanto's Lustran ABS. The furniture guard strip is PVC. Made by Hoover Ltd.

f **HAIRDRYER**

The casing is ABS; the component that holds the motor is glass-reinforced polypropylene. Courtesy ICI Plastics Division.

a **DUSTPAN AND BRUSH**

The dustpan is polystyrene with a rubber lip. The brush is polypropylene. Courtesy Vileda Ltd.

b **BATH BRUSH**

b, c, d **TOILET AND BATHROOM ACCESSORIES**

The brushes are nylon with polypropylene backs and holder. Courtesy Vileda Ltd.

a

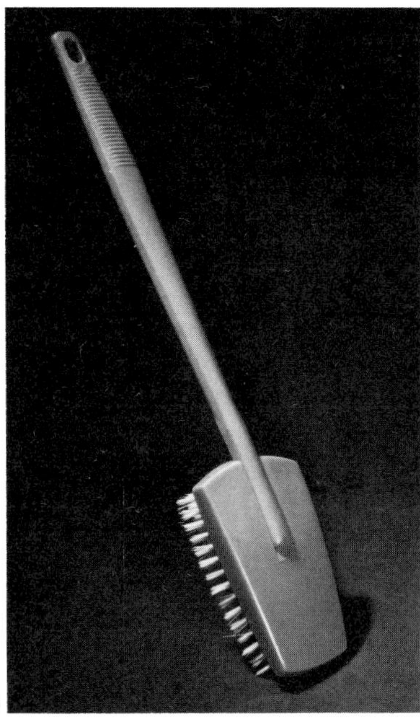

b

c

d

JEWELLERY & SCULPTURE

GUITAR

Made out of clear 'Perspex' by Guy Davies, this is a full-size working instrument, the first of its kind. An electric guitar's body is non-functional, giving scope for an interesting sculpted form. Courtesy ICI Plastics Division.

a **NECKLACE**

Silver and resin. The resin infill is transparent black, translucent off-white and bright opaque green. Designed and made by Susanna Heron. Courtesy Crafts Council.

b, c **BIRD RINGS**

Also by Susanna Heron. Courtesy Crafts Council.

a

b

c

d **BANGLE AND RING**

Both of perspex. Designed and made by Nuala Jamison. Photo David Ward. Courtesy Crafts Council.

e **BROOCH**

Aluminium, acrylic and polyester resin. Designed and made by Eric. Photo David Cripps. Courtesy Crafts Council.

f **BRACELET**

Designed and made by Anne Finlay. Photo David Ward. Courtesy Crafts Council.

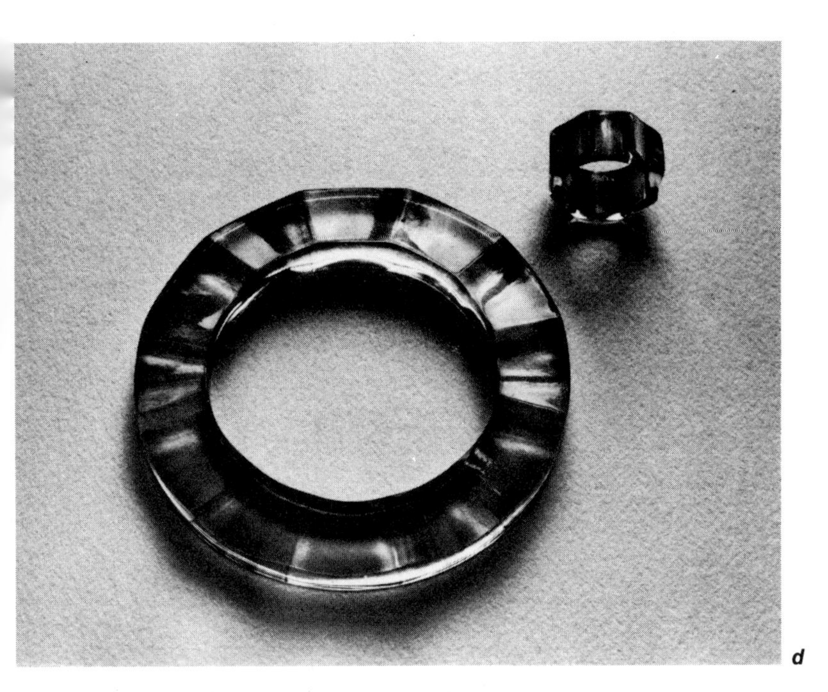

d

e

f

a **TORSION**

Perspex sculpture by Naum Gabo. Courtesy Tate Gallery, London.

b **SPIRAL THEME**

Also by Gabo. Courtesy Tate Gallery, London.

a

b